Gaslighting in America

A Pictorial Survey, 1815-1910

Denys Peter Myers

Dover Publications, Inc., New York

Published in Canada by General Publishing Company, Ltd., 30 Lesmill Road,
Don Mills, Toronto, Ontario.
Published in the United Kingdom by Constable and Company, Ltd., 3 The
Lanchesters, 162–164 Fulham Palace Road, London W6 9ER.

This Dover edition, first published in 1990, is an unabridged, corrected republica-
tion of the work first published under the title *Gaslighting in America: A Guide for
Historic Preservation* ("Heritage Conservation and Recreation Service Publication
Number 3") by the U.S. Department of the Interior, Heritage Conservation and
Recreation Service, Office of Archeology and Historic Preservation, Technical
Preservation Services Division, Washington, D.C., 1978. For the Dover edition the
author has written a new Preface (replacing the old Foreword) and corrected a
number of typographical errors in the original text.

Manufactured in the United States of America
Dover Publications, Inc., 31 East 2nd Street, Mineola, N.Y. 11501

Library of Congress Cataloging-in-Publication Data

Myers, Denys Peter, 1916–
     Gaslighting in America : a pictorial survey, 1815–1910 / by Denys Peter Myers.
          p.   cm.
     Reprint. Originally published: Washington, D.C. : U.S. Dept. of the Interior,
Heritage Conservation and Recreation Service, 1978.
     Includes bibliographical references and index.
     ISBN 0-486-26482-3
     1. Gas-lighting—United States—History—19th century.   2. Gas-lighting
—United States—History—20th century.   I. Title.
TH7960.M93   1990
621.32′4—dc20
                                                                                    90-46149
                                                                                       CIP

Table of Contents

Preface to the Dover Edition

Gaslighting in America: A Pictorial Survey, 1815-1910 (with the original subtitle *A Guide for Historic Preservation*) was first published by the U.S. Department of the Interior in 1978. The book was planned as one of a series of reports on technical aspects of historic preservation in response to Executive Order 11593, signed on May 13, 1971, under which the Secretary of the Interior was given the responsibility for developing and disseminating "information . . . for preserving, improving, restoring and maintaining historic properties." Lee H. Nelson, FAIA, until recently Chief, Preservation Assistance Division, Technical Preservation Services, National Park Service, and Sarah M. Sweetser Theis, Architectural Historian, worked with the author to edit the manuscript, make a final selection from his plethora of illustrations and design the book. The first and only printing was soon exhausted, and the book has remained out of print until this reprint by Dover Publications, Inc.

Since this book first appeared in 1978, several gas-fixture catalogs not mentioned in the first edition have come to the author's attention. A circa-1876 Mitchell, Vance & Co. catalog containing 83 plates with over 1,000 illustrations of every variety of gas fixture, fitting and glass shade then offered by that leading New York City firm is now readily available in a Dover reprint under the title *Picture Book of Authentic Mid-Victorian Gas-Lighting Fixtures* (Dover 0-486-24640-X).

A visit to the Essex Institute in Salem, Massachusetts, to inspect the splendid Henry N. Hooper & Co. catalog of 1858, which contains 62 lithographed plates of elegant "Chandeliers, Girandoles, Candelabra, and Lamps, in great variety of pattern, and of the most approved styles, for Gas, Oil and Candles, finished in Plain, Olive and Antique Bronze, Ormolu, or Gold and Silver Plate . . .," will amply reward the interested student. The library at Winterthur contains some 36 American trade catalogs related to lighting, ranging in date from 1858 to 1924. Among the gas-fixture catalogs, the circa-1871 Mitchell, Vance & Co. *Catalogue of Crystal Glass Chandeliers, Brackets, Standards, &c.* is outstanding and would merit a trip to Winterthur all by itself. The remarkable collection of material illustrating Victorian decorative arts now at the Athenaeum of Philadelphia includes much pertinent material, among which is an undated Mitchell, Vance Company catalog of combination gas and electric fixtures. While in Philadelphia, the student should not neglect to visit the Historical Society of Pennsylvania for a perusal of the circa-1876 Cornelius & Sons catalog, three plates from which are illustrated herein.

While there is nothing so instructive as the examination of extant unrestored original fixtures, certainly the necessary supplement is the study of contemporary catalogs. The beautifully lithographed Archer, Warner, Miskey & Co. and Archer & Pancoast Mfg. Co. catalogs in the Metropolitan Museum of Art, and the McKenney & Waterbury Co. catalog in the Avery Library at Columbia University, samplings from which are reproduced in the following pages, are major inducements to visit New York City. The same applies, of course, to Old Sturbridge Village, Massachusetts, which possesses the fascinating Starr, Fellows & Co. catalog, seven plates from which are reproduced herein. In Washington, the 69 engravings of circa 1820-30 representing the wares of an unknown British manufacturer of lighting fixtures certainly warrant a visit to the Library of Congress, and the 41 photographs assembled in 1882 by Thackera, Sons & Co. should lead the interested researcher to the National Archives.

Since *Gaslighting in America* was first published, additional data have come to light. It now appears that the gas table lamp illustrated on Plate 32 may have been one of a pair affixed to a mantelpiece shelf, like those in the former sitting room of the John Tucker Daland House of 1851-52, now part of the Essex Institute in Salem, Massachusetts. Certain information has been kindly furnished by colleagues. Research done by Betty Bird in 1984 for the U.S. Treasury Department shows that the Cash Room chandeliers mentioned on page 139 were actually supplied by M. L. Curtis & Co. of New York City, although the Tucker Manufacturing Company supplied other gas fixtures for architect Mullett's north wing. Professor Francis R. Kowsky of the State University of New York at Buffalo has written to say that Gothic gas standards similar to those shown on Plate 80 were designed by architect Frederick Clark Withers (1828-1901) for his First Presbyterian Church of 1859 in Newburgh, New York, and manufactured by Cornelius and Baker. Unfortunately, the standards no longer exist. Perry Gerard Fisher, formerly Executive Director-Librarian of the Columbia Historical Society, wrote in 1979 to inform the author that the gas and electric chandelier illustrated on Plate 109 was made in Vienna, Austria.

Denys Peter Myers
Consulting Architectural Historian

BOSTON & SANDWICH GLASS

Acknowledgements

It is a pleasure to thank the many people without whose kind assistance the writing of this paper would have been onerous indeed. John C. Poppeliers, Chief of the Historic American Buildings Survey, graciously acceded to requests for specially taken photographs, and Jack E. Boucher, HABS Photographer, devoted much overtime and skill to supplying illustrative material. Henry A. Judd, Chief Historical Architect, National Park Service, kindly supplied data on Ford's Theater.

James M. Goode, Curator of the Smithsonian Building, was exceedingly generous both with his time and with photographs of gas fixtures in the collection under his care, none of which had been previously photographed. My thanks go to him and Rodris Roth, Curator, Division of Costumes and Furnishings at the Smithsonian, for their thoughtful review of the manuscript. Donald J. Lehman supplied much information on pertinent material in the National Archives and on the artist-designer Richard von Ezdorf. William Seale lent a number of particularly informative photographs from his extensive collection of interior views. Keir Helberg spent a day showing his restored gaslighting in Baltimore and discussing the fine points of various types of burners. Craig Littlewood offered useful data on early manufactures.

Suzanne Boorsch of The Metropolitan Museum of Art's Print Room, Etta Falkner, Librarian of Old Sturbridge Village, Edith Nisbet, Librarian of the American Gas Association, Arlington, Virginia, and Wendy Shadwell of the New York Historical Society were helpful well beyond the call of duty. David Sellin, Curator of the U.S. Capitol, and Florian H. Thayn in the Architect of the Capitol's office assisted materially with research of obscure questions regarding the correspondence of the two Philadelphia firms of Cornelius and Baker and Archer and Warner relating to the Capitol.

Mary Ison and C. Ford Peatross of the Prints and Photographs Division of the Library of Congress were helpful and took a personal interest in the research. Stanley Brown, Douglas Helms, and Michael Musack of the National Archives were also most generous with assistance of more than a routine kind. The Bernard Fensterwalds of Alexandria, Virginia, allowed photographs to be taken of one of their chandeliers. Last, but certainly not least, thanks are due an Alexandria resident who allowed her gaslighted chandelier to be photographed but who wishes to remain anonymous. In the course of a prolonged research project, it is almost inevitable that some who helped are overlooked. To them go both thanks and an apology for the omission of their names. It is also almost inevitable in a pioneering research effort that factual errors occur. For whatever errors may have crept in, the author alone is responsible.

Acknowledgement for the Dover Edition

The author and publisher gratefully acknowledge the cooperation of Lee H. Nelson, FAIA, until recently Chief, Preservation Assistance Division, and staff, U.S. Department of the Interior, National Park Service, Preservation Assistance Division, Technical Preservation Services, Washington, D.C., in the preparation of the Dover edition of this book.

Frontispiece — View of Boston and Sandwich Glass Company Exhibit, 1878.

Overall, American gas fixtures ranged from the lavishly elegant to the starkly plain. This half of a stereograph shows the Boston and Sandwich Glass Company's display at the annual exhibition in Mechanics' Hall, Boston, 1878. It illustrates not only typical American glass chandeliers and wall brackets (commonly miscalled "crystal") among the most elaborate gas fixtures of the period, but also at the upper right a fixture of the simplest form, an iron pipe gas "T."

This strictly utilitarian "T" (which was not part of the exhibit) has six burners; the more typical examples had only two. Also illustrated on the glass chandeliers are the then recently introduced wide-necked glass shades on slender wire supports called "spiders," and the hanging triangular notched spear prisms that first appeared around 1850.

Courtesy of the Sandwich Historical Society, Sandwich, Massachusetts.

List of Plates

Frontispiece: View of Boston and Sandwich Glass Company Exhibit, 1878.

Introduction

This report focuses on the types and styles of gas fixtures which appeared in the rooms and on the streets of 19th and early 20th century America. It does not describe the scientific methodology for the manufacture of gas, nor the technology of pipe installation.

A chronological approach has been adopted in the hope that the text and illustrations may serve as guides for avoiding anachronisms in preservation projects, as well as exemplifying the types of installation appropriate to specific situations. To this end, we have included many kinds of fixtures known to have been used to a considerable extent in public or private buildings and streets, ranging from the simple iron "T" to the elaborate crystal chandelier. Certain devices that were seldom used in America (if the contemporary pictorial evidence is a reliable guide) have not been discussed or illustrated. The emphasis has been on the norm, not the exceptions to almost universal practice.

Background

As with most scientific discoveries, no individual can be credited with the "invention" of gaslighting. As early as 1739 in London, John Clayton (1693-1773) reported the results of his "Experiment Concerning the Spirit of Coals" in the *Philosophical Transactions* published by the Royal Society. Clayton successfully distilled gas from burning coal but made no economically practical application of his discovery. During the rest of the 18th century, various experimenters demonstrated gaslighting on a limited scale without significant results.

At the turn of the 19th century, another Britisher, William Murdock (1754-1839), raised gaslighting from the status of a curiosity to a practical alternative to candles and lamps. He accomplished this feat by distributing coal gas by pipes to light the Boulton and Watt Soho Works at Birmingham in 1798 and the Phillips and Lee factory at Salford in 1805. Perhaps even more significant than the success of the apparatus was the fact that the Salford factory was lighted by gas for a cost of £600 compared to £2,000 a year for candles. In 1808, the British scientific community recognized Murdock's achievement by awarding him the Royal Society's Rumford Medal "for a treatise on his application of the illuminating properties of carburated hydrogen [coal gas] for the purpose of furnishing a new and economical light."

In the meantime, the French inventor Philippe Lebon (1767-1804) was granted a patent by the First Consul Bonaparte in 1803 for distilling illuminating gas from wood. Lebon's work attracted the attention of an enterprising and persistent German entrepreneur named Friedrich Albrecht Winzer (1763-1830), who acquired the Lebon patent. In 1804 Winzer traveled to England, where he Anglicized his name to Frederick Albert Winsor. He abandoned the use of wood gas for coal gas and set about organizing a company for the manufacture and distribution of the new illuminant. Winsor lighted an extensive segment of Pall Mall in London by gaslamps on June 4, 1805. He had business organizing abilities that Murdock lacked, and in spite of the latter's opposition, Winsor obtained a charter in 1812 from Parliament for the first gaslight company — The London and Westminster Chartered Gas Light and Coke Company. Thereafter, with the invaluable aid of a notable engineer named Samuel Clegg (1781-1861) and others, the company flourished.

Gaslighting spread sporadically to other urban centers and then to smaller communities. It spread more rapidly in Great Britain and in the United States than on the European continent, but it eventually became the predominant 19th century illuminant in heavily settled areas throughout what was then called the "civilized" world. In the last two decades of the century, gaslighting was challenged by electric lighting, but gas remained popular for street lighting until the outbreak of the First World War in 1914. Specific data on the growth and chronology of American gas companies by geographic location are given in the text and notes accompanying plate 111 and in the Appendix of this report.

Gaslighting Fixtures

The following illustrations and commentary cannot, of course, answer every question that may arise, but it is hoped that at least the main outlines of American gaslighting practice have been clearly drawn. To give an example of how modern concepts can subtly influence decisions erroneously, it was recently considered most appropriate to hang a crystal gas chandelier in the dining room of a *circa* 1860 house. Contemporary evidence indicates, however, that crystal, or glass fixtures were almost never used in dining rooms, although they occurred with some frequency in parlors. Another instance of misunderstanding, in a restoration purporting to represent the 1860s, was the use of gas mantles which were not developed until the late 1880s. This report provides guidelines to prevent anachronisms and the misapplication of styles of fixtures, their burners and finishes.

In researching gas fixtures, it is important to examine actual fixtures, as well as illustrative material of the period. First-hand examination of the lighting devices in this manner by the restoration artisan improves the quality of the restoration, and prevents two major errors. First is the misinterpretation of scale from either a distorted photograph, or an inaccurately drawn sketch. Second, is the failure to reproduce the correct finish on the metal and glass components of the fixture. Old lithographs, engravings, and photographs rarely suggest the correct color or finish. Only a careful examination of original fixtures that have not been improperly refinished, can serve as a true guide. The precise colors and shadings of brass, ormolu, "bronze" finishes, frosted glass, and so on, can best be achieved by carefully copying fixtures of the period to be represented. Finally, it should be remembered that it is preferable to use original fixtures rather than reproductions, provided that the suitable fixtures can be found.

Exterior lighting has its own history and restoration problems. Because of numerous exterior photographs and streetscapes available from the 1860s to the present, determining appropriate style is generally not difficult. Stylistic changes did not occur as frequently with exterior lighting; in many areas the same lampposts remained despite modernizing improvements to the lantern. However, reference to pictorial evidence from the same general location, time period, type of building or street scene is important. Furnishing the gas for exterior lighting involves many of the same problems as for interior fixtures, such as complying with municipal codes.

Was Gaslighting Used?

Before making any selection of gas fixtures, physical or documentary evidence must be present to indicate that gaslighting was used in the building or street under consideration. Capped gas butts in areas such as back corridors, attics, cellars, or butler's pantries may provide visible evidence. If no visible evidence exists, gas pipes within walls can supply convincing evidence, provided that they can be shown not to have serviced heating or cooking appliances rather than lighting fixtures. Because gas pipes were run within the walls and did not show, it is highly improbable that they were ever entirely removed when electric lighting was installed in a formerly gaslighted building. Therefore, if there is no physical trace of gaslight system within a structure, any documentary evidence seemingly to the contrary should be interpreted with the greatest caution. Having established by physical evidence or by a combination of physical and documentary evidence that the structure was once gaslighted, it could be ascertained by research in local archives whether the gas was supplied from a gas company or from a private gas machine.

Connecting the Gas

The use of gas is rarely a concern for preservationists, having been precluded by its scarcity, the increased restrictions of state and local building codes and, more recently, by the national concern for conservation of fossil fuels. For the few cases where gas lines and service are still intact, the following factors should be taken into account.

The gas now available is not the manufactured coal gas used almost universally in the past. It is natural gas, which burns with a more intense heat than coal gas. Beyond that, there are certain dangers to be considered. Coal gas leaks were easily detectable by smell, whereas natural gas requires the use of an additive to give it a discernible odor. It was recently found that natural gas leaking into the fill near a foundation loses its warning odor and can travel along buried pipe into the cellar with explosive results. Under no circumstances should plastic pipe be used, because it has a tendency to split, permitting the undetected escape of gas. All pipe must be metal, and the completed piping installation must undergo stringent testing for leaks, as set forth in Gerhard's *American Practice of Gas Piping* (see Bibliography). Repairing existing pipe will require cutting into walls, floors and ceilings.

Where an ornamental plaster medallion indicates the previous existence of a ceiling fixture, the restored fixture should be piped by raising portions of the flooring above, and laying the pipe between the joists. In a few instances, the upper flooring will be ornamental (i.e. parquet) and should not be disturbed, in which case, careful "excavation" of the plaster ceiling may be necessary. Chandeliers should be supported by iron brackets secured to adjacent ceiling joists and not supported by pipe alone. It is also desirable to support all but the lightest wall brackets by iron braces secured to adjoining wall studs or by utilizing expansion joints in masonry mortar joints.

Electrification

However, when it is not feasible to reinstall gas piping, electric light can be regulated with the use of low voltage transformers or silicon controlled rectifier "dimmers" to simulate the quality of gaslight. Despite the fact that electricity cannot convincingly imitate open flame lighting, which flickers occasionally, the color and level of light can be achieved.

Until the introduction of the incandescent gas mantle, or Welsbach burner (*circa* 1890), gas burners gave relatively little light. The average flat flame burner, whether of the fishtail (union jet) or batswing type, did not deliver a maximum of over 16 candlepower under optimum conditions. Even the Argand burner, which was little used in America except for the center burners of slide fixtures (see plates 83, 84, and 95) and for some "portables," or lamps, gave only a slightly greater amount of light. Therefore, if electrical simulation of gaslighting is decided upon, dimmers or transformers should be provided. Candle socket bulbs should be used as the larger diameter of standard sockets destroys the desired illusion of gas burnertips. Candle sockets are actually larger than gas tips as well, but their size is more sympathetic in diameter to the real burners. The plastic sleeves with which electric candle fixtures are supplied should be painted dark gun metal grey, unless the fixtures originally had gas burners designed to imitate candles (shown in plate 15). The bulbs used to imitate flat flame jets should not exceed ten or fifteen watts at the most. A ten-watt glass flame bulb emits the amount of light needed to simulate a lighted fishtail burner. There are also bulbs available commercially that imitate gas mantles satisfactorily. It is difficult, dazzled as we are by the intensity of modern lighting, to realize how comparatively little light satisfied our forebears, but it is as important to restore the correct color and level of light as it is to restore the fixture.

The history of gaslighting styles provides an interesting and sometimes amusing commentary on the needs and decorative aspirations of our 19th century citizenry. Some of the fixtures pictured herein were unique, others were commonplace to everyday life. Understanding the scope of the devices available, their typical uses, and their appropriateness to the historic needs of the project at hand will complement the numerous efforts required for the preservation of the fabric and character of historic structures today.

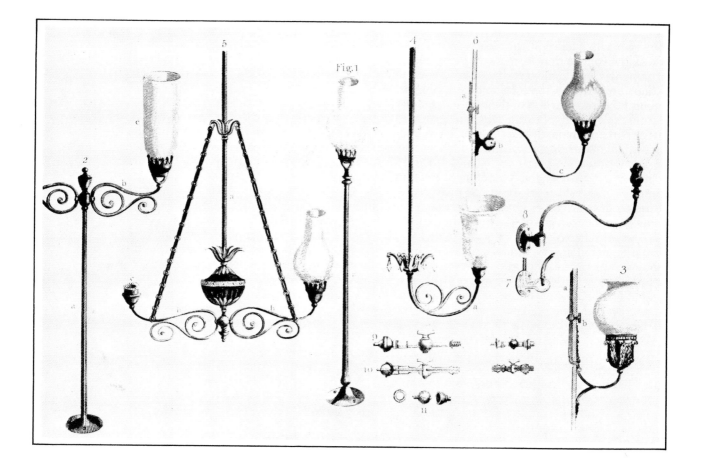

No record of gas fixtures made in the United States before the late 1830s has been found. Until ca. 1840, it appears that gas fixtures used in America were imported from England and, to some extent, France.

The earliest illustrations of gas fixtures are probably those published in 1815 by Rudolph Ackermann, the London art dealer and publisher of books and prints, as embellishments of Fredrick Accum's *Practical Treatise on Gas-Light.* Represented here are fixtures "already in use in this Metropolis." [1] From left to right, they are described as 2) "Rod Gas Lamp with branches," 5) "Pendent Double-Bracket Lamp," 1) "Rod Lamp," 4) "Pendent Rod Lamp," 6) "Swing Bracket Lamp," 3) "Bracket Lamp." Except for figure 8 (right center), a "Swing Cockspur Lamp," the burners shown appear to be of the Argand type. It is worth noting that the terms "bracket" and "pendent" persisted in use, although "rod" types were soon referred to as "pillar."

Also around 1815, Ackermann published an aquatint by J. Bluck after a watercolor by Augustus Pugin (not reproduced here) showing possibly the earliest view of a gaslighted interior, Ackermann's Art Library.

Reprinted from Fredrick Accum, *A Practical Treatise on Gas-Light,* London, 1815, Plate 3, courtesy of American Gas Association, Inc. Library, Arlington, Virginia.

Ornamental fixtures in Accum's treatise on gaslighting, London, 1815.

Plate 2

It is sometimes fallaciously thought that early gas fixtures were simple, even primitive, in form and that ornamental fixtures are therefore comparatively late in date. This plate from Accum's *Practical Treatise on Gas-Light* of 1815 clearly proves otherwise. "The gas-lamps exhibited in this plate, are employed in the library, counting-house, warehouse, and offices of Mr. Ackermann . . ."[3] On this plate, figures 1 through 8 are described respectively as "a Candelabrum, an Arabesque Chandelier, a Roman Chandelier, a Gothic Chandelier, a Pedestal Figure Lamp, a Pedestal Vase Lamp, a Girandole, and a Candelabrum." Observe that as early as 1815, the eclectic taste so characteristic of the 19th century already embraced "Arabesque, Roman, and Gothic" designs. With the exception of the cockspur at the top of figure 1, all the burners are of the early type termed "rat-tail" (see plate 3). These fixtures were finished in greenish-bronze and highlighted with gilding.

Reprinted from Fredrick Accum, *A Practical Treatise on Gas-Light*, London, 1815, courtesy of American Gas Association, Inc. Library, Arlington, Virginia.

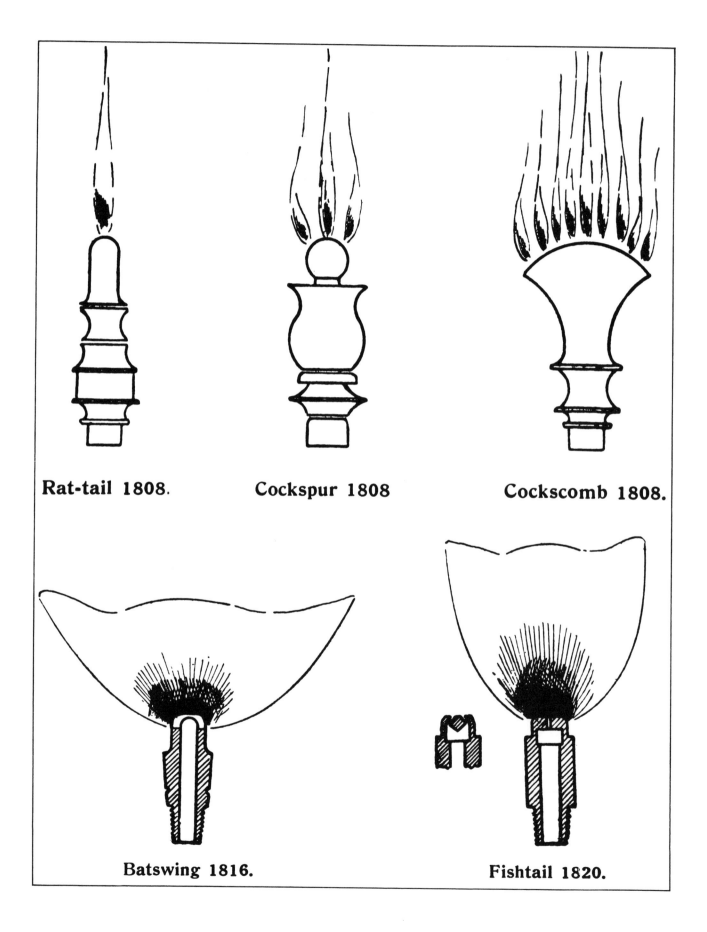

Rat-tail 1808.   **Cockspur 1808**   **Cockscomb 1808.**

Batswing 1816.   **Fishtail 1820.**

Five of the six types of burners in use by ca. 1820 are shown here, the sixth in plate 4. The cockspur and the cockscomb were wasteful and inefficient and soon fell into disuse. The rat-tail continued in use primarily in a modified form designed for burners imitating candles. The batswing burner, used for street lamps and other outdoor illumination, and the fishtail burner were, except for gas candles, the almost universally used forms of burners until the introduction of the Welsbach mantle for general use in 1890.[4] The batswing burner had a domical top pierced by a narrow slit across it. The fishtail, or union jet, burner was apparently invented by the Scotsmen James Neilson and James Milne.[5] It was so designed that two jets of equal size impinged on each other to produce a flat flame issuing from a single small aperture. The tops of fishtail burners were usually slightly concave and were pierced by a small central hole.

Very short lengths of pipe with either fishtail, batswing, or rat-tail burners were termed "scotch tips."

Reprinted from Dean Chandler, *Outline of History of Lighting by Gas*, London, 1936, courtesy of American Gas Association, Inc. Library, Arlington, Virginia.

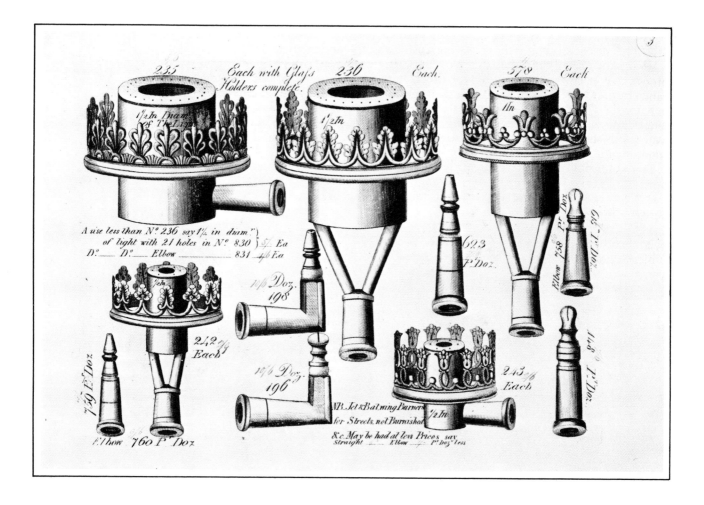

235   Each with Glass   236   Each.   578   Each

Holders complete.

1/2 In Diameter of The Light

1/2 In

1 In

A size less than Nº 236 say 1¼ in diamʳ) of light with 21 holes in Nº 830 } 5/6 Eª

Dº ___ Dº ___ Elbow ___ 831 __ 4/6 Eª

623 Pʳ Doz.

Elbow 758 Pʳ Doz.

757 Pʳ Doz.

10/6 Doz. 198

7 In

212 2/9 Each

759 Pʳ Doz.

10/6 Doz. 196

Elbow 760 Pʳ Doz.

1/2 In

243 2/6 Each

NB. Jet & Bat wing Burners for Streets, not Burnished &c May be had at less Prices say

Straight ___ Elbow ___ Pʳ Doz. less

748 Pʳ Doz.

This illustration and the following eight plates are from a series of 69 unidentified British engravings of lighting fixtures dating around 1820-1830.[6] The burners with the ornamental galleries, multiple holes, and hollow centers operated on the Argand principle, producing a circular column of flame with air at the center and around the periphery. The Argand gas burner, the sixth type in use early in the development of gaslighting, does not appear to have been used extensively in the United States, although it was popular in England for interior illumination as late as the 1840s. It required the use of a glass chimney, whose drawbacks were frequent breakage and the constant need for cleaning. The crown-like galleries shown here were supports for these chimneys.

The burners with the slitted tops are batswing; those with the single holes, here termed "jet" burners, are probably fishtail, or union jet burners. Early burners, whether of iron or brass, frequently clogged because of corrosive impurities in the gas. Suppliers furnished small augers, narrow slips of brass, and small saws to clear burners. After the introduction of the steatite, or "lava," burner tip in 1858, clogging became less of a problem. These noncorroding tips, invented by M. Schwarz of Nuremburg, were made of a variety of Bavarian soapstone which had been subjected to slowly increased heat and subsequent boiling in oil. Schwarz was granted an American patent on July 20, 1858.[7]

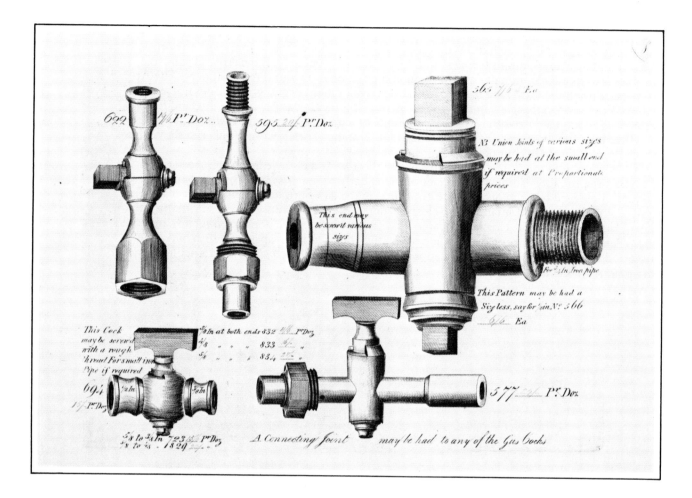

622   1/6 P.r Doz..

595 20/ P.r Doz

565 7/6 Ea

N.o Union Joints of various sizes
may be had at the small end
if required at Proportionate
prices

This end may
be screw't various
sizes

For 3/4 In Iron pipe

This Pattern may be had a
Size less, say for 1/2 in N.o 566
6/- Ea

This Cock
may be screw'd
with a rough
thread For small iron
Pipe if required

694   1/2 In

1/ P.r Doz

3/8 In at both ends 832   24/ P.r Doz
1/2   "   "   833   26/ "
5/8   "   "   834   28/ "

3/8 In

5/8 to 3/4 In 723 34/ P.r Doz,
5/8 to 7/8 . 1829 24/ "

577 24/ P.r Doz

A Connecting Joint   may be had to any of the Gas Cocks

This plate from the before-mentioned series of unidentified British engravings shows clearly the pin and partial collar safety device which prevented the gas cocks from being turned too far and unintentionally left on. After it was discovered that the pins provided insufficient security against accidental breakage and consequent asphyxiation, threaded cocks that precluded all possibility of leakage were frequently used.

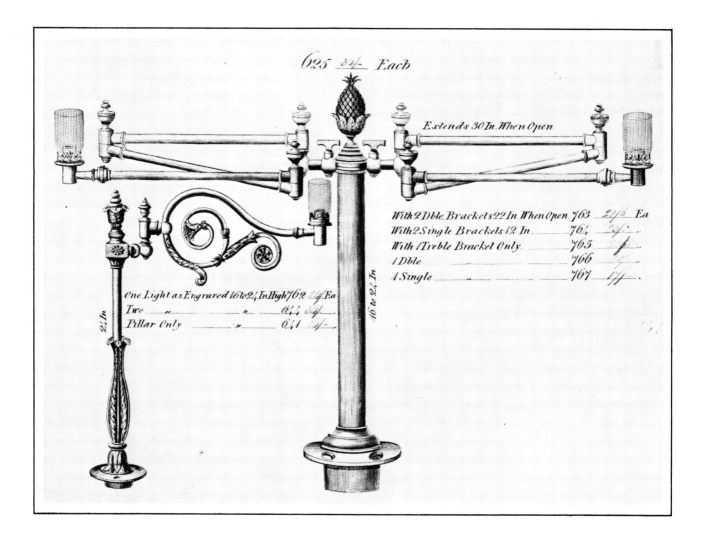

625 34/- Each

Extends 30 In. When Open

With 2 Dble. Brackets 22 In. When Open. 763 29/6 Ea
With 2 Single Brackets 12 In. 764 29/-
With 1 Treble Bracket Only. 765 29/-
1 Dble 766
1 Single 767 4/-

One Light as Engraved 16 to 24 In High 762 24/ Ea
Two " " 644 34/-
Pillar Only " 641 4/-

24 In.

16 to 24 In.

This plate shows that so-called "pillar" fixtures and, more significantly, jointed branches were already in use during the 1820s. Jointed branches continued in use until the end of the gas era, particularly in bedroom fixtures. Their flexibility was particularly advantageous where light was desired in close proximity to mirrors. Jointed branches were also used near desks, or wherever light was needed for close work.

Most of the burners shown in the unidentified British engravings dating from ca. 1820-1830 are of the Argand type, and have the characteristic straight, tubular chimneys necessary for the Argand burner, but do not have shades. A manuscript page preceding the series notes that "articles may be had bronzed to order at the same price as lacquered," indicating that the principal finish was probably burnished brass, which required lacquer to prevent tarnishing.

From the Library of Congress, Prints and Photographs Division, Lot 2728.

699   12 In 5/ Pr Pair

74   Iron   30 Inches   6/   Each

12 Inches

700
4/ Each

May be had to Extend 24 Inches
With Glass Icicles &c
No. 740 26/ Each

639

16 Inches
2/ Pr Pt

Th29hese wall brackets, like the chandeliers illustrated on plate 2, show both the elaborateness of early gas fixtures and the eclectic stylistic character of their design. The uppermost bracket had traces of Baroque influence, although the foliate motifs were neoclassical in manner. The second bracket was based on architectural elements of the Gothic style rather than on any actual medieval prototype. Note that this fixture was iron and was probably gilded, whereas the other branches were probably brass. The serpentine bracket was typical of Regency taste in its least classical and most fanciful manifestation. Fantastic winged creatures appeared as late as 1856 in an American catalogue on bracket branches, and a bracket was designed in the form of a rattlesnake as late as 1859.[8] The foliate bracket at the bottom of the plate was throughly Greek Revival in the anthemion (honeysuckle) motif. The ball joints of all these brackets indicate that they were designed to swing from side to side.

753 may be had any number
of Lights at 21/- Pr Light extra

18 In Bracket to match 753
is 840 42/- Pr Pair

1 Light Pendant to match
753 is 841 27/- each
With Joint Top 1/- ea extra

753 85/- each

36 In between the Lights

May be had on a less
Scale say 28 In between
Lights
754
74/- each

"Pendant gaselier" from unidentified English catalogue, ca. 1820-1830.

Plate 8

The term "pendant" used in the text of this plate was, by the 1840s, applied only to fixtures having one or two lights. "Chandelier" or "gaselier" (also spelled "gasolier" or "gasalier") were used interchangeably for fixtures having three or more lights until the 1860s, when "chandelier" ultimately prevailed in common usage.

The slack chains of the pendant shown here were ornamental, not functional. Such chains were frequently used to decorate gas fixtures until the mid-1850s and were characteristic of gaselier design until about 1850. This 1820s example is severly neoclassical in design compared with fixtures of the late 1840s and 1850s. Note that more branches could be added and that the span of 36 inches could be reduced to about 28 inches.

From the Library of Congress, Prints and Photographs Division, Lot 2728.

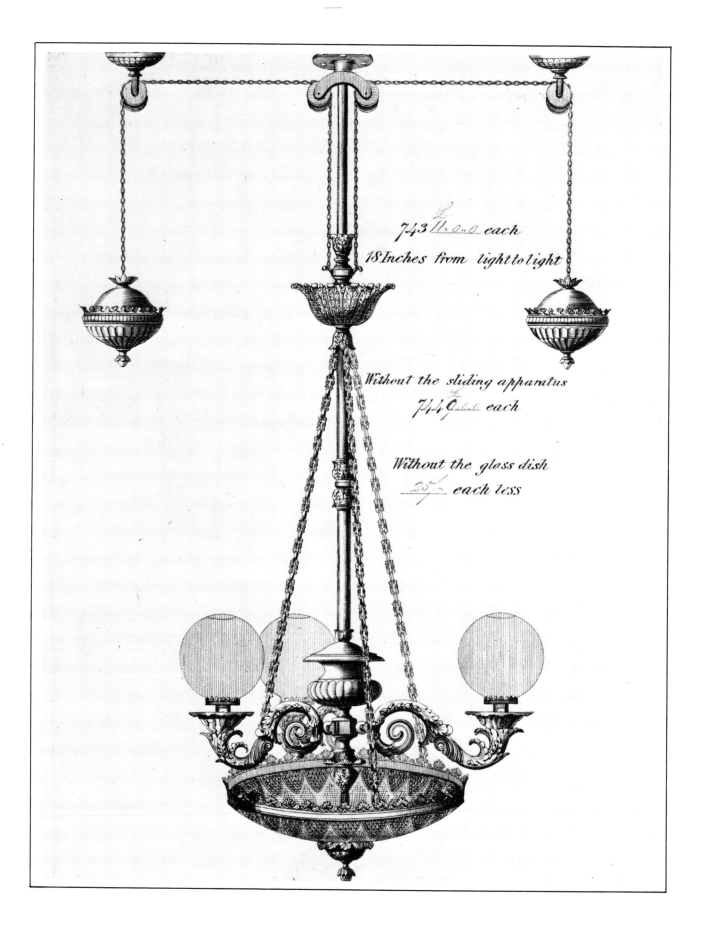

743 £ 11.0.0 each

18 Inches from light to light

Without the sliding apparatus
744 £ 0 each

Without the gloss dish
25/ each less

Water seal gaselier from unidentified English catalogue, ca. 1820-1830.

Plate 9

This plate provides evidence that water-seal gaseliers, which could be raised or lowered, were available as early as the 1820s. Gas had several advantages over oil lamps and candles, among them greater safety from fire, less smoke, and no grease or oil spills. But it had one real disadvantage — lack of portability. To overcome that disadvantage, numerous ingenious devices were designed such as the jointed extensible branches depicted on plate 6; and here the water seal at the top of the outer sliding stem permitted the raising or lowering of a gaselier without danger of gas leakage. A film of oil prevented rapid evaporation of the water.

In later examples, the counterweights were suspended from pulleys attached directly to the outer stem, omitting the extra set of pulleys fastened to the ceiling as shown in this plate.

Movable suspension chandeliers such as this were frequently used where light was desired close to a table. Hence, they were often used in dining rooms or over library or parlor center tables. However, there was no strict rule regarding their use. There is at least one documented instance of the use of chandeliers over the aisle, where there were no tables, between the bar and the dining booths in a St. Louis oyster saloon.[9]

From the Library of Congress, Prints and Photographs Division, Lot 2728.

27

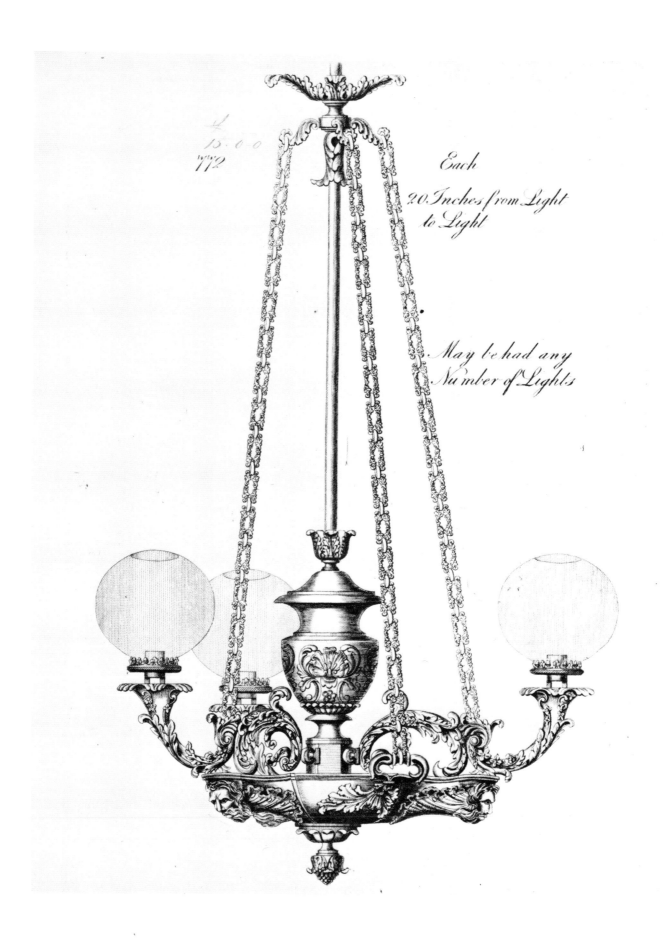

772

Each
20 Inches from Light
to Light

May be had any
Number of Lights

Plate 10

Three-branch chandelier from unidentified English catalogue, ca. 1820-1830.

Chandeliers similar to the one shown here were installed in the library of Sir Walter Scott's "Abbotsford" near Edinburgh by 1823.[10] Different patterns of this general form, i.e., pyramidal shape with a single tier of lamps, remained popular into the 1850s. Later examples substituted pipe rods for the chains and omitted the center stem.

This particular style is Neo-Renaissance more than neoclassical. Note the male masks on the soffit of the base, or "bowl"; both male and female heads, half figures, and figures, as well as animal forms were favorite ornamental motifs until after 1860.

The "urn" and "bowl" in this example are holdovers from oil chandelier designs and serve no function in the operation of a gas fixture.

The globes of this chandelier and that on the previous plate suggest that fishtail burners may have been used, yet the large circular forms of the burners themselves indicate Argands. If Argand burners were used, the engraver omitted the requisite chimneys, perhaps in the interest of simplification.

From the Library of Congress, Prints and Photographs Division, Lot 2728.

29

529

£11.0.0 Each

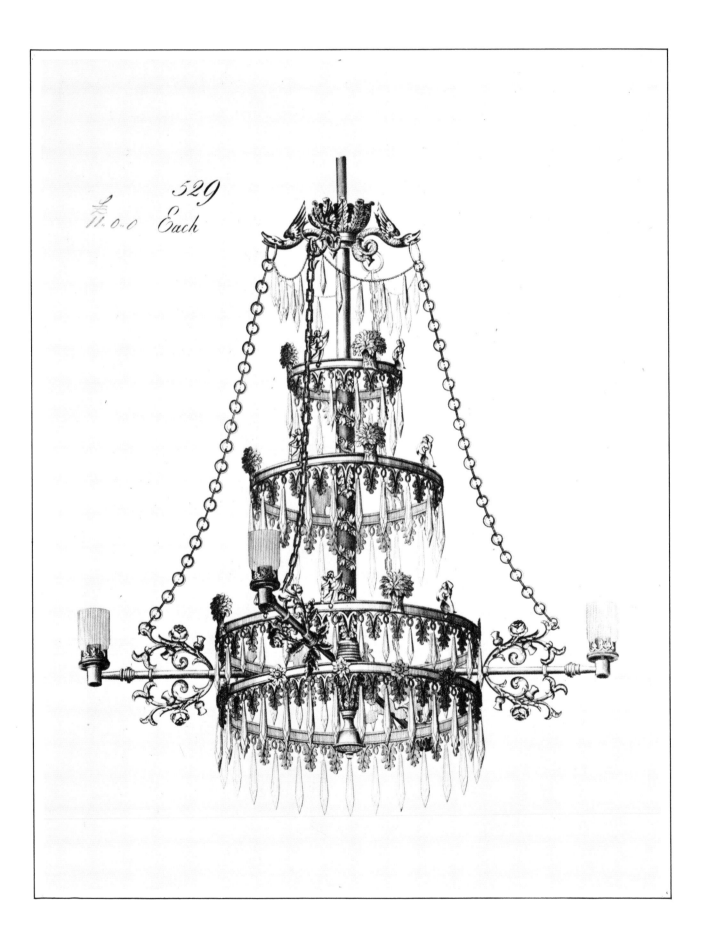

Chandeliers ornamented with ascending concentric prism-hung rings originated during the English Regency and passed out of fashion during the 1840s. The prisms were aptly termed "icicles."

Later examples of the ring-supported prism type show much more glass in proportion to the metal visible, and the chains seen here do not appear on later examples. The supporting elements of some American examples dating from the 1840s were silvered. However, the usual finish of the metal parts was gilt ormolu.

The rather naturalistic ornament of the four branches of this example, uniting the English rose and the Scottish thistle, was most certainly British; but the small dancing figures suggest possible French Empire influence. The serendipitous expression of the mythological creatures at the top seems to indicate that they found the chains unexpectedly delicious!

A Boston ballroom, ca. 1838.

Plate 12

In a New Englander's memoirs, the author recalls:

> Gas was not introduced into dwelling-houses [in Boston] until Pemberton Square
> was built by the Lowells, Jacksons, and their friends, in the years 1835, 1836, and
> later. It was a surprise to everyone when Papanti introduced it in his new Papanti's
> Hall. To prepare for that occasion the ground-glass shades had a little rouge
> shaken about in the interior, that the white gaslight might not be too unfavorable
> to the complexion of the beauties below.[11]

These same beauties may be those in this undated lithograph by B. W. Thayer from a music
cover titled "Tremont Quadrilles." In 1838, the noted Boston dancing master Lorenzo Papanti
opened a ballroom in Tremont Row on Tremont Street that was "lighted by crystal chandeliers
imported from Paris at a cost of $1,200."[12] The costumes, hair fashion and the lighting indicate
that this may well be the occasion depicted here. It does not, as has sometimes been thought,
illustrate a ballroom in the famous Tremont House, a Boston hotel built in 1829.[13]

Note that because no oil fonts are visible, the fixtures are assuredly gas—the chandeliers,
brackets, and, on the musicians gallery parapet, gas "pillars." This is one of the earliest views
of a gaslighted American interior.

Reprinted from music cover, "Tremont Quadrilles," B. W. Thayer and Company. Lithograph courtesy of the American
Antiquarian Society, Worcester, Massachusetts.

¶In his "Philosophy of Furniture" (1840) Edgar Allen Poe wrote:

> We are violently enamored of gas and of glass. The former is totally inadmissable within doors. Its harsh and unsteady light offends . . . The huge and unmeaning glass chandeliers, prism-cut, gas-lighted and without shade, which dangle in our most fashionable drawing-rooms, may be cited as the quintessence of all that is false in taste or preposterous in folly.[14]

The "huge glass chandeliers, prism-cut, gas-lighted, and without shade," shown in this half of an 1866 stereoscopic view, "dangled" in the most prominent (if not the most fashionable) drawing-room in America—the East Room of the White House. Despite their offending Poe's aesthetic sensibility, there seems little doubt that President James K. Polk liked the fixtures, because he had the candle-holding chandeliers converted to burn gas in 1848, by the noteworthy Philadelphia firm of Cornelius and Company who supplied new gas fixtures for other major rooms of the White House as well. The three East Room chandeliers had been bought by President Andrew Jackson from the Philadelphia upholstering firm, Lewis Veron and Company, on June 7, 1834, at a total cost of $3,300.[15] In 1873 these chandeliers were replaced by President Grant with the even more elaborate "crystal" fixtures shown in plate 70.

Whatever the scale of the fixture, burners varied only slightly in size. Because of problems related to gas pressure, the method of obtaining more light was to increase the number of burners per chandelier rather than to enlarge the size of the individual burner. There appear to have been 26 burners on each of these fixtures.

In 1841, George Albert Lewis's parents moved to a new house at the corner of Walnut and 16th streets in Philadelphia. Many years later Lewis painted from memory this watercolor of the third story front chamber that he and his bride occupied in the house from 1851 to 1855.[16]

The gas pendant depicted in the view probably dates from 1841, as the neoclassical treatment of the two branches and, particularly, the arrangement of tiers of prisms, are typical of many chandeliers equipped with either gas burners, candles, or Argand lamps from the 1820s through the 1840s. The prisms of this bronze-finished chandelier were of the flattened triangular type, popular after about 1840. As the stems of chandeliers were, like the shafts of lamps, often called "pillars," the prisms used to decorate this portion of a fixture were referred to as "pillar icicles" at that time. In 1834, the New England Glass Company of Cambridge, Massachusetts, had produced the first "pillar icicles" made in America.[17] By 1846, pillar icicles arranged in tiers adorned the chandeliers in the ladies' saloon of the first vessel to be illuminated by gas, the Long Island Sound steamboat *Atlantic*.[18] An excellent example of this type hangs in the Office of the Senate Minority Leader in the U.S. Capitol. It has eight branches and five tiers of prisms.

The Lewis watercolor provides important evidence that suspended gas lamps were used at least occasionally in the 1850s, if not earlier. In this instance, the lamp appears to be attached to a pipe extension rather than to a hose.[19] In any case, the shade, apparently made either of green silk or of fluted paper, cannot have offered great security against the accident of fire.

PLATE XVII

Gas Fittings &c.
Executed by
Cornelius and Son
Philadelphia.

Cornelius "Gas Fittings," 1846, from T. U. Walter and J. J. Smith, <u>A Guide to Workers in Metal and Stone.</u>

Plate 15

This 1846 print is the earliest illustration of gas fixtures known to date that can be definitely ascribed to American manufacture.[20] John Henry Frederick Sachse, one of the leading artist-designers in the Cornelius establishment during the 1840s, may well have designed the fixtures on this plate.

In the caption "Gas Fittings &c/Executed by/Cornelius and Son/Philadelphia," the "&c" refers to the oil-burning lamp and candelabrum at the lower left and lower right respectively. What appear to be candles in the chandelier and bracket are gas burners masked by ceramic or opaque glass sleeves to resemble candles.[21] The lavishly ornamented style of these Neo-Renaissance lighting fixtures prefigures the rage for luxurious splendor (equated in the popular mind with rich ornamentation) that prevailed during the 1850s.

The Cornelius firm was preeminent among American manufacturers of gas fixtures for many years.[22] Christian Cornelius, the founder of the firm, emigrated from Amsterdam in 1783.[23] He established himself in Philadelphia as a silversmith and, later, as a manufacturer of plated ware.[24] By the mid-1820s, the firm was engaged in making lamps and chandeliers, and by 1833 silverplating was no longer mentioned in the *Philadelphia Directory* listing.[25] On February 8, 1836, the Philadelphia Gas Company began operations[26] "and as soon as the use of gas was introduced, the firm began to turn their attention to supplying the necessary appliances for its consumption."[27] In 1831 Robert Cornelius, (1809-1893) joined his father in partnership. From then until Christian Cornelius's death in 1851, the firm was known as either "Cornelius and Son" or "Cornelius and Company."[28] The career of this important firm after 1850 will be noted later.

Robert Cornelius, who headed the firm from 1851 until 1876, studied chemistry under the distinguished chemist and geologist, Professor Gerard Troost (1776-1850), and drawing under the noted Philadelphia drawing master, James Cox (1751-1834).[29] He had a decidedly scientific and mechanical bent and perfected devices for facilitating his manufacturing processes. In 1841 he secured the first American patent for an improvement to a gas jet. He also invented and patented a solar lamp for burning lard or sperm oil. His interest in chemistry led him to develop improved processes for electroplating, and he was among the first to use bromine in daguerreotypy, thereby reducing the required exposure time from ten minutes to ten seconds. In October or early November 1839, Robert Cornelius was one of the first people ever to photograph a human countenance — his own.[30]

Reprinted from Thomas U [stick] Walter, and J. Jay Smith, *A Guide to Workers in Metal and Stone. . . .* Philadelphia, Carey and Hart, 1846, Plate XVII. Library of Congress.

This Sarony and Major lithograph recorded the interior of the Broadway Tabernacle in New York City during the distribution of the American Art Union prizes in 1847. In contrast to the ornate examples by Cornelius and Son, this illustration shows that many gas fixtures used in the 1840s were simple in design.

The unadorned iron corona suspended from the dome is equipped with short burners of the type called "Scotch tips." The pillar lights flanking the pulpit and the brackets on the gallery parapet are almost equally modest. The corona is not unlike a greatly enlarged version of the gas ring used by Rembrandt Peale at his Museum in Baltimore, when he gave the first public demonstration of gaslighting in America on April 23, 1816.[31] Later the same year, gaslighting was also demonstrated at Charles Willson Peale's museum on the second floor of Independence Hall in Philadelphia. Before the Broadway Tabernacle (built in 1836) was demolished in 1857, an elaborate chandelier had replaced the corona.

The use of such unpretentious gas rings with "Scotch tips" evidently persisted for some years. A wood engraving of Chickering's Hall in Boston, made in 1869, shows a fixture similar to the one in the 1847 view. In that interior the elaborate architectural treatment of the hall and the elegance of its other lighting fixtures, contrast oddly with the gas ring. Chickering Hall was fitted with brackets each having a pair of shaded branches with clusters of gas candles above. Possibly the gas ring was a temporary fixture added to give extra light for the rehearsal shown in the wood engraving.[32]

Reprinted from *Distribution of the American Art-Union Prizes, Broadway Tabernacle, 1847, 340 Broadway near Worth Street*, a Sarony and Major lithograph published by John P. Ridner, courtesy of The J. Clarence Davies Collection, Museum of the City of New York.

Richard Caton Woodville's painting titled "Politics in an Oyster House" dates from 1848. The artist clearly shows the plain accoutrements of the modest establishment, including the severely simple iron pipe gas elbow on the rear wall of the booth. Such fixtures, and iron pipe gas T's like the one with Scotch tips shown in the frontispiece, were the plainest models made. They continued in use until the end of the gas era both in very inexpensive and unpretentious buildings and in starkly utilitarian structures such as prisons, mills, manufactories, and even some hospitals and schools.

Chandelier with Morning-Glory Patterned Branches, ca 1850.

Plate 18

The floral and foliate ornaments of this chandelier are typical of the mid-19th century fondness for natural forms treated in a manner derived, often somewhat distantly, from 18th-century Rococo precedents.

The particular pattern of the branches, based on the morning-glory vine, was evidently very popular, if one may judge by the number of surviving examples. Formerly, at least four chandeliers with branches of this same casting hung in Quarters One at Springfield Armory in Massachusetts, and there was at least one of the type in the Moses Myers House in Norfolk, Virginia.[33] Except for two minor details, one of which is merely an inversion of a small decorative element, the two parlor chandeliers once at Springfield Armory and a pair of chandeliers formerly in the John Duval Howard House at 209 West Monument Street in Baltimore and now in the Maryland Historical Society's Thomas and Hugg Room were identical.[34] Two other museums, the Daughters of the American Revolution Museum in Washington, D. C., and the Missouri Historical Society, each have a chandelier with branches of this same morning-glory pattern.[35] Another excellent example is in the parlor of the Vassal-Craigie-Longfellow House in Cambridge, Massachusetts. The James T. Johnston House in Alexandria, Virginia, has a four-light example of this pattern and once had matching wall brackets. The Ebenezer Maxwell Mansion in Philadelphia possesses a fine pair of two-light brackets with branches of the morning-glory pattern.

One last example, the parlor chandelier of the Thomas Jefferson Southard House in Richmond, Maine, should be cited, especially for the remarkable finish of its spun brass parts. They are "damasked" in a floral pattern, blue on gold, like the blade of a presentation sword. A contemporary account of the manufacture of such fixtures records:

> Some of the ornamental work is painted in parti-colors to please fanciful tastes; some is bronzed with different shades; while other work is tastefully enameled or covered with a coating of fine gold. There are also rooms . . . appropriated to the workers in artistic bronze, while others are occupied by those who are employed at *damask* work, in which the chief agents are lacquer and acids.[36]

When found in Baltimore a few years ago, this chandelier was painted white. The fact that paint remover did not damage it attests to the excellence of the original finish, now restored. Even a black and white photograph shows the contrast between the gloss of the lacquered spun brass parts and the matte gilding of the cast ornaments and branches. Such contrasts played a major part in the effect produced by brass chandeliers of the 1840s and 1850s. It is incorrect to refinish or produce them in one tone of gilt without any interplay between bright and matte surfaces.

This type of chandelier had either four or six branches of the morning-glory pattern and ornamental vertical chains. However, several of them differ noticeably in detail, giving evidence that the parts were interchangeable, thus allowing the manufacturer or local gasfitters to assemble to suit individual orders. It is significant that the often cited principle of interchangeable parts, first applied in 1800 by Eli Whitney to the manufacture of firearms, was also applied to the manufacture of gas fixtures. As noted by the Cornelius and Baker firm:

> All the screws of the different classes that are turned out of this establishment are made of one size. If the branch of a chandelier exported by this house to China should find its way to Russia, it would fit exactly into any of the chandeliers in the Kremlin.[37]

The four ornamental chains once supplied with this chandelier have been lost. The spiral sleeve masking the main gas pipe is a modern reproduction, as well as the gas keys and supports for the shades. The original keys which have survived on the example at the Missouri Historical Society (mentioned earlier) are larger. On the other hand, the original shades had necks of a much smaller diameter. Shades with throats the diameter of those shown here, 3½ inches, were not made until after 1876.

From the author's collection, photograph by Jack E. Boucher.

45

This detail of the chandelier shown in the previous plate reveals the basic structure to which the ornamental elements were added. The brass parts of the fully assembled chandelier were supported by plain iron gas pipe, as seen here. (The holes in the pipe were made by an impatient electrician threading wire through the fixture). The button-like object between the branch in the foreground and that at the right is a plug sealing the outlet for a threaded jet to which the hose of a gas table lamp could be attached. All the cast ornaments of this chandelier are attached by screws to the spun brass parts.

Because these branches match the branch castings on some of the fixtures in Quarters One at Springfield Armory, a tentative attribution of this fixture might be made to the New York firm of Starr, Fellows and Company.

However, it must be noted that the catalog, dated 1856, states that the firm began making gas fixtures "only about six years since" (e.g. 1850) whereas the house at the Springfield Armory was fitted with gas at its completion in 1846.[38]

From the author's collection, photograph by Jack E. Boucher.

47

This chandelier, ornamented by a blue-over-clear cased glass bowl and baluster vase of the type popularly called "Bohemian," has well chased gilt bronze castings of excellent quality. It dates from around 1850, was used in Massachusetts, and was probably made in Boston. The chandelier has been tentatively attributed to Henry N. Hooper and Company, with the glass parts attributed to either the Boston and Sandwich Glass Company or the New England Glass Company.[39]

Central glass elements are very unusual. In 1848, frosted and etched glass bowls and vases of similar or identical shape were used on a pair of parlor chandeliers in the Valentine-Fuller House in Cambridge, Massachusetts. That evidence supports a Boston provenance for this chandelier. But the attribution to the Hooper firm must remain putative at best, as several manufacturers of gas fixtures were active in Boston at the time, among them G.D. Jarves and Cormerais, and Henry B. Stanwood and Company.[40] It is interesting to note that Stanwood advertised his firm as "manufacturers and importers."

It should be mentioned that the Cambridge Gas Light Company was not formally chartered until 1852, though it had commenced operations a few years before its charter was granted. As early as 1848, the Valentine-Fuller House and, in 1850, the George Washington Whittemore House had city gas in Cambridge. A number of American gas companies were in business before receiving their charters.[41] In Washington, D. C., for example, a functioning gas company existed as early as 1848, but it was not chartered until 1855.

Courtesy of The Metropolitan Museum of Art, Rogers Fund, 1969.

Archer and Warner grapevine patterned gaselier, 1850.

Plate 21

This Neo-Rococo six-branched gaselier was manufactured by the Philadelphia firm of Archer and Warner and was included in their catalogue as shown on plate 23. The treatment of the grapevine motif is naturalistic and is typical of mid-century fashions in design. Ornamentation derived from the grapevine was also popular as a carved motif on furniture of the period.

The fixture was originally fitted with globe shaded burners rather than gas candles. Later it was slightly shortened and electrified.

Archer and Warner competed with Cornelius and Baker in the quality of their work, though not in the volume of their production. Ellis S. Archer, a Philadelphia merchant, patented a lard lamp on June 8, 1842, and thereafter engaged in its manufacture. In 1848, Redwood F. Warner joined Archer in partnership, and by 1850 the firm had become a leading manufacturer of gas fixtures, lamps, and girandoles. On March 18, 1850, patents were issued to them on designs for brackets and chandeliers, one of which was this dated example.[42] One Archer and Warner design for a bracket was based on the fuchsia plant and was analogous in spirit to the morning-glory branches seen in plates 18 and 19.[43]

Courtesy of the Reverend and Mrs. W. B. Morton III, Waterford, Virginia, photograph by Jack E. Boucher.

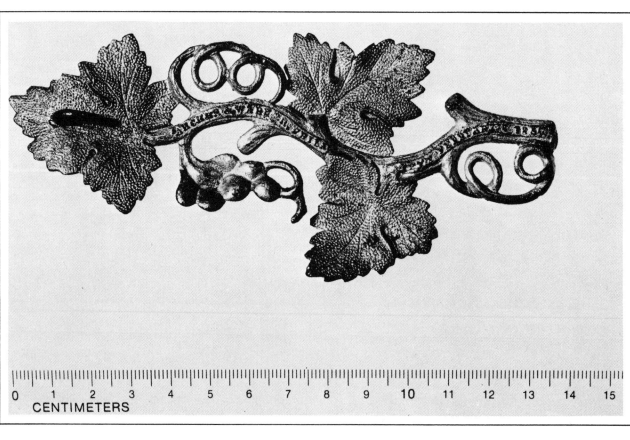

If at all labeled, most chandeliers carry their maker's mark on the gas turn keys. The Archer and Warner example in plate 21 has highly ornamental keys in the form of grape leaves and bunches of grapes precluding placement of a maker's mark. Hence, the unusual positioning of the mark illustrated here is on the reverse side of each link of the decorative chains. Barely legible here is the patent date "March 19, 1850."

Patents were often sought as much for their advertising value as for any other reason. Indeed, Archer's lard lamp of 1842 appears not to have been a very practicable bit of gadgetry but patented primarily for the publicity value.[44] However, the Archer and Warner design patents of March 19, 1850, may have been sought in response to a pirating of their designs by minor manufacturers. There is some evidence that unscrupulous makers actually recast their competitors' designs from molds taken of the original fixtures.[45]

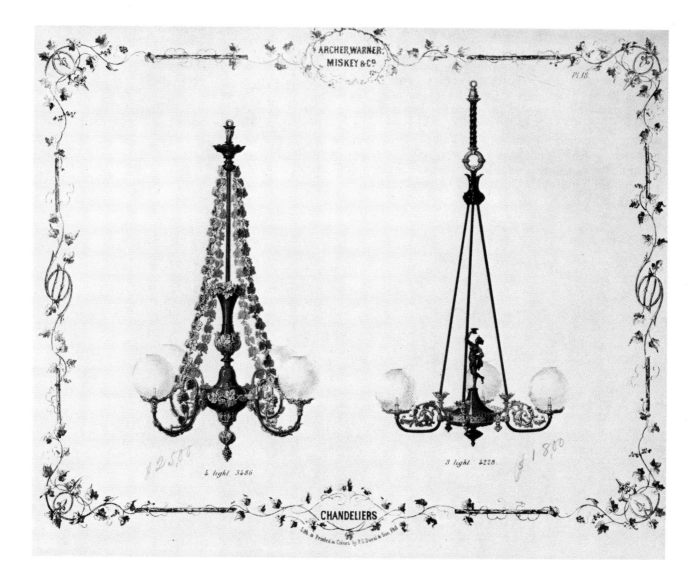

ARCHER, WARNER, MISKEY & Cº

Pl.18.

4 light 3486

3 light 4228

CHANDELIERS

Lith. & Printed in Colors by P.S. Duval & Son Phil.ᵃ

Plate from Archer, Warner, Miskey and Company catalogue, 1857-1859, showing gaselier with grapevine motif.

Plate 23

T his page from an Archer, Warner, Miskey and Company catalogue advertises the same chandelier illustrated in plate 21. The chains, branches, and ornamental finial below the bowl are identical. The vase ornament which is not present on the extant chandelier may have been discarded. Also, the bowl is a variation and lacks its cover. As already noted, the actual fixture has been slightly shortened and electrified.

The colors of this plate are green, yellow, and orange, indicating bronze, burnished gilt, and matte gilt. The globular shades were evidently unfrosted, as the burners are visible through them.

Rococo chandelier, ca. 1850.

Plate **24**

Originally from an Alexandria, Virginia, building of the early 1850s, this chandelier depends more upon formalized Rococo treatment of foliate motifs and a little less upon naturalistic renditions of natural forms than the examples shown on plates 18, 20, and 21. The branches of those three preceding chandeliers are formed of paired castings joined together, whereas the branches here are made of brass tubing to which the ornament has been applied. The foliate crown above the chains, the bobeches below the shades, the "spiders" (shade holders), and the plaster ceiling ornament are modern, as well as the brass canopy at the top. Such canopies are required for the installation of electrified fixtures to protect the wires, which are usually joined with wire nuts at that point. This chandelier has been shortened about eight inches. When found, bronze powder and banana oil radiator paint had totally obscured the original finish. Cleaning has revealed the contrast of matte and burnished gilding and bright lacquered brass proper to its appearance. The globes, frosted with the popular griffin pattern, date from around 1880 and postdate the chandelier itself by about 30 years.

When electrifying a gas fixture the shortest possible candle fittings should be used, unless gas candles were originally present. While the diameter of even the smallest electric candles is too great to imitate gas burners with complete success, shades will normally help to conceal this defect. Painting the sleeves of the electric candles a dark gold, to resemble tarnished brass, or gun metal or a matte black, to resemble iron burners, will enhance the effect. Unless dimmers are used, clear glass bulbs of no more than 10 watts should be used to approximate the light emitted by a standard fishtail burner.

From the author's collection, photograph by Jack E. Boucher.

Parlor gaselier, ca. 1856, still functioning with gas burners.

Plate **25**

This relatively simple gilt ormolu and lacquered brass gaselier of a charmingly light and delicate vine pattern is a considerable rarity. It is one of a matched pair *in situ* in a large parlor and has never been electrified. They are in an Alexandria, Virginia, house that was completed in 1856. The light visible through the turn-of-the-century shades comes from burners actually burning gas. Compare the diameter of the burners, visible through the later shade-holders, with the diameter of the electric candles in the preceding plate. Note also that no canopy is used; the gas pipe goes directly into the ceiling through the rich, original plaster centerpiece. Many early chandeliers have later shades, as in numerous cases the original shades were broken. Often they were replaced by widemouthed shades, around 1880 or later, to increase the efficiency of combustion by admitting more air to the burners (which also helped to eliminate flickering). As the keys of this chandelier and the one in plate 24 are identical, it is safe to assume that both fixtures are by the same unidentified maker.

From a private collection, photograph by Jack E. Boucher.

Allegorical gaselier, ca. 1853.

Plate **26**

Allegorical bronze statuettes were very much *en vogue* during the 1850s as ornaments for gaseliers. The three females gracing the stem of this fixture from the 1853 Edmund Ira Richards House (now demolished) in North Attleboro, Massachusetts, represent art, science, and industry.[46] The same figures appear on the two parlor chandeliers of the 1851 Robert Campbell House in St. Louis, and on the parlor chandelier of "Camden," the Pratt family seat completed in 1859, in Caroline County, Virginia.[47] Another six-branched chandelier, now in the Smithsonian Institution Castle, has a trio of somewhat more sprightly statuettes—short-skirted female figures perhaps representing Flora, Pomona, or other minor deities.

The branches and other components of the Campbell House and "Camden" fixtures differ from those illustrated here. But the Smithsonian example has branches identical with these, and the hall chandelier of the Wickham-Valentine House in Richmond, Virginia, has the same keys as both this one and the one in the Smithsonian. It therefore seems probable that these last three fixtures were made by the same firm, possibly Cornelius and Baker.

It should be noted that none of the chandeliers mentioned here had the ornamental chains so characteristic of the 1840s. The shades of this chandelier have the small-necked bases appropriate for the date of the fixture. But their globular form and the Neo-Grec character of their acid-etched banding could not date earlier than the 1860s. The canopy above the stem suggests *Art Nouveau* and appears to date from around 1900.

Courtesy of The Metropolitan Museum of Art, Gift of Mrs. Frederick Wildman, 1964.

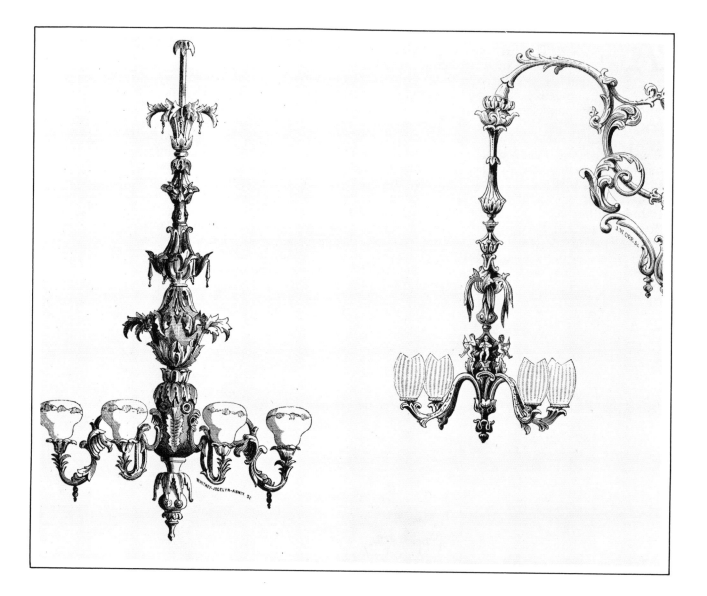

Engraving of Cornelius and Baker fixtures shown at the International Exhibition in New York City, 1853-1854.

Plate 27

At the Great Exhibition of 1851 in London's Crystal Palace, Cornelius and Baker showed a pair of richly ornamented, lacquered brass 15-burner gaseliers measuring about 15½ feet high by 6 feet wide. The keys represented "bunches of fruit, thus combining beauty with utility."[48] Two years later the firm exhibited at least three gas fixtures, including the two shown here, at the New York Crystal Palace during the International Exhibition of 1853-1854. The third was a four-light bracket of heavily gilded bronze. A putto springing from acanthus *rinceaux* held the branches, only two of which were engraved. Apparently one key opened all four branches just below their springing point.[49]

The two fixtures represented in this wood engraving were described as follows:

> The Chandelier is very rich and beautiful, suited to adorn as well as illuminate the apartment in which it may be hung. The bronze of which it is made has a tint of rich, deep green, which is relieved with admirable effect by the brilliancy of the gilding applied to the decorative parts. The adjoining Bracket with a pendent chandelier of four lights is also characterized by elegance of form and ornamentation, excepting, however, the little figures perched just above the branches of the lights. These have no adaptation to a chandelier, and violate a fundamental law of decorative art, that all ornamentation should rise out of construction and belong to it.[50]

Chandeliers depending from large brackets, like the ensemble at the right, were often called "toilets" when placed next to dressing mirrors. At least one fine set of such pendants cantilevered on brackets still exists, projecting from the gallery parapets of the Beneficent Congregational Meeting House ("Round Top Church") in Providence, Rhode Island.

As noted previously, upon the death of Christian Cornelius in 1851, his son Robert and his son-in-law Isaac F. Baker formed a partnership.[51] Within three years William C. Baker also became a partner. Whereas Isaac F. Baker was listed as "lamp manufacturer," William C. Baker was listed as "merchant."[52] In 1859 Robert's son, Robert Comeley Cornelius, became a partner, and by 1861 another son, John C. Cornelius, was a partner in Cornelius, Baker and Company.[53] Evidently, Robert Cornelius was the driving force behind the success of the company from the time he joined his father in 1831. By 1857 the company had two factories, linked by telegraph, in operation, one at 181 Cherry Street and the other at Columbia Avenue and 5th Street in Philadelphia.[54]

Elaborate chandelier, probably by Cornelius, ca 1849.

Plate 28

This richly ornamented gilt bronze and lacquered brass fixture hangs in the Ashburton House, now St. John's Parish House, in Washington, D.C. It probably dates from 1849, when the house was reacquired from an interim owner by Her Majesty's Government for use again as the British Legation.[55] The chandelier is particularly well populated, as there are six putti emerging from the foliate scrolls of the branches, three angels on the vase above the bowl, and three ladies standing symbolically on the stem. Unfortunately, the quality of the modern shade reproductions accords poorly with the superlative workmanship of the fixture.

Although it is unmarked, a strong case can be made stylistically for attributing this chandelier to the Cornelius firm (cf. plate 27). If the putative date of 1849 is correct, there was no other firm in America at the time, with the possible exception of Archer and Warner or Henry N. Hooper, capable of producing work of this quality.

Courtesy of St. John's Episcopal Church, Washington, D.C., photograph by Jack E. Boucher.

Light has traditionally accompanied festivity, and this wood engraving of a ball (for the officers of the Imperial Russian Atlantic Fleet at the Academy of Music in New York City on November 5, 1863) shows that extra light was sometimes provided by temporary pipes rimming the horseshoe-shaped galleries of the auditorium with rows of burners. Curiously, the half of the engraving shown here depicts the burners at the base of the first tier parapet as shaded, where the left half of the print (not shown) represents Scotch tips without shades.[56] The large chandeliers within the stage were also installed for the ball.

The lighting of theater auditoria varied. The New York Academy of Music, built in 1854, was normally lighted by multibranched brackets affixed to the gallery parapets. Its domed ceiling had no chandelier. The Opera House at Niblo's Garden in New York had a similar arrangement of lighting fixtures. However, the very large Boston Theatre in 1854 had an auditorium lighted by an immense prism-hung chandelier. The 1863 Ford's Theatre in Washington, now restored, has a six-branched chandelier suspended in front of each set of proscenium boxes and a row of single-branched brackets on the second balcony ("family circle") parapet.

One great mid-century American theatre, the Academy of Music of 1857 in Philadelphia, has survived substantially unchanged. Cornelius, Baker and Company supplied both gallery parapet brackets and a notable prism-ornamented central chandelier to light the auditorium. The chandelier originally had 240 gas burners and measured 16 feet wide by 25 feet high. It was "said to be the largest in the world."[57]

CHARLES OAKFORD'S MODEL HAT STORE,

·····▷· 158, Chesnut Street, Philadelphia ·◁·····

HATS, CAPS & FURS, WHOLESALE & RETAIL.

Standards, or "pillars," were used with some frequency for lighting shop counters and bars during the 1850s. This Duval lithograph clearly shows the lighting of Charles Oakford's Model Hat Store in Philadelphia. The elaborate fittings, including the three-light gas pillars, were installed in 1854. Oakford so prized his fittings that when he moved from this store to the Continental Hotel in 1860 he reused everything movable except the marble floor.[58] The Boston jewelry store of Jones, Ball and Company had similar lighting, except there the three-light pillars were posed on pedestals set between the glass topped counters.[59] A lithograph of the Gem Saloon in New York titled "Temperance, but no Maine Law" published by A. Fay in 1854 shows the marble bar and mirror-topped back bar with single light pillars.[60] The interior of Thomas Brothers' Bar Room as shown in a wood engraving published in the New York *Illustrated News* for October 6, 1860, was lighted by chandeliers and by seven-branched putto-supported pillars on the bar.[61]

The lighting of stores varied greatly. A Rosenthal lithograph of L. J. Levy and Company's dry goods emporium of 1857 in Philadelphia shows elaborate pillars, brackets, and chandeliers.[62] A wood engraving of the much less elaborate dry goods establishment of James Beck and Company in New York shows rather simple four-branched chandeliers and, over the counters, a large number of single-burner harp fixtures of the type commonly used to light front entries. Peterson and Humphry's carpet store in New York was lighted by numerous gas T's. Gas T's were also used to light Cushings and Bailey's bookstore in Baltimore, but there Argand burners were substituted for the more usual fishtail burners to provide ample reading light.[63]

GREAT FAIR GIVEN AT THE CITY ASSEMBLY ROOMS, NEW YORK, DECEMBER, 18

More often than not, burners were protected from drafts by glass shades. However, shades were not invariably used. This detail from a wood engraving after a drawing by Winslow Homer shows the forms of typical fishtail jet burners unshaded at the tips of the chandeliers' branches. This illustration was published in *Harper's Weekly* for December 28, 1861, and represents the interior of the City Assembly Rooms, built ca. 1859, in New York City during a charity fair in late 1861.[64]

As early as 1853, gas table lamps were in use. A watercolor of Mrs. A. W. Smith's parlor at Broad and Spruce Streets in Philadelphia painted in 1853 by Joseph Shoemaker Russell shows a gas table lamp attached by a slender brass elbow to a bracket. The lamp is a quite simple one, composed of a spirally turned standard on what appears to be a small square marble base.[65]

This gilt-bronze example, one of a pair by an unidentified maker, has been dated ca. 1855 and was used in Baltimore.[66] As the gas is supplied by a pipe from underneath the base, its form is unusual for a gaslamp and is related to the fixed pillar concept. The more commonly used portable single-burner lamps of the period were fed gas through a rubber hose connected at the side of the base. The elaborate Neo-Rococo design combines flowers, fruit, foliage, a putto with a bird (at the springing point of the arms), and four small heads, two male and two female, on the arms. The glass bobeches and teardrop pendants add further glitter to an already eye-filling *tour de force* of lavish ornament.

The shades are of a type normally found on oil lamps rather than on gas burners. They certainly are not well adapted for fishtail jets and could not have worked at all with batswing burners. No other instance of the use of shades of this shape with gas burners is known on American fixtures.

Courtesy of The Metropolitan Museum of Art, Rogers Fund, 1967.

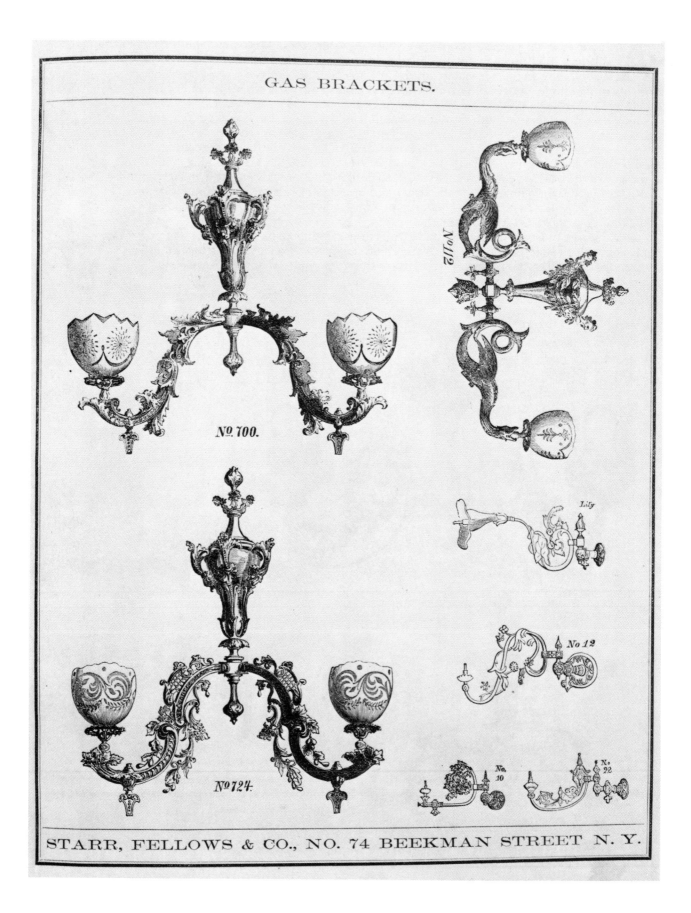

No. 700.

No. 112.

Lily

No. 724.

No 12.

No. 10.

No. 22.

STARR, FELLOWS & CO., NO. 74 BEEKMAN STREET N. Y.

Until after the Civil War, Philadelphia retained its lead as the principal manufacturing center of American gas fixtures primarily because of the volume produced in that city by Cornelius, Baker and Company. New York and Boston were also producing significant quantities of gas fixtures during the 1850s and later (see plate 20). For example, New York City directories recorded two gas fixture makers in 1847, but by 1860 their number had risen to twenty-four.[67] Not the least important among New York firms was Starr, Fellows and Company, which became Fellows, Hoffman and Company on February 1, 1857.[68] Fortunately, a copy of a Starr, Fellows catalogue dated 1856 with Fellows, Hoffman addenda through 1858-1859 has survived. This ephemeral publication is the earliest complete American catalogue of gas fixtures that has been rediscovered.[69] It is an invaluable source of information for the positive identification of many unmarked fixtures.

The preface to the catalogue states that "a few samples only" are presented, as the "styles are continually changing." The text added, "We get up, to order, Chandeliers, Pendants, and Brackets, of any design, as regards pattern, or size that may be desired." Furthermore, "These are all made in the various colors of Gilt, Olive, French Bronze, Artistic Bronze, or any two colors in combination and with or without Slides."[70] The lithographed illustrations were made from drawings done by young ladies who were students at the New York School of Art. The firm cited this as evidence that they were patronizing a worthy cause; however, it is possible that the budding artists may have been paid somewhat less than the rate required by full-fledged professionals. The text reference to this and the preceding page of gas brackets (not reproduced) reads as follows: "These Brackets are finished in any modern Fancy Color desired. Nos. 40, 734, and 317, on this and 700, 724, and 112 on succeeding page [shown here] make very fine Church Fixtures." It is difficult to perceive any specifically ecclesiastical character in their design, however. Salesmanship sometimes takes odd turns!

Morning-Glory bracket by Starr, Fellows and Company, 1856.

Plate **34**

This is a morning-glory bracket (mislabeled "Lily") shown on the preceding Starr, Fellows and Company plate. Fixtures with glass "blossoms" of this type imitating morning-glories, lilies, or fuchsias enjoyed a minor vogue during the 1850s. Characteristically, the gas light jetted straight out instead of upward as on the more conventional models. This and a similar bracket, probably by another maker, are in the collection at the Henry Ford Museum. Plate 1 of the catalogue of the Archer, Warner and Miskey designs (not shown) illustrates another "Lily" among their collection of swing brackets.

Courtesy of the Collections of Greenfield Village and the Henry Ford Museum, Dearborn, Michigan.

Another item attributable to Starr, Fellows and Company through plate 33 of this report is this chandelier, whose branches are identical with those of the bracket numbered 700. However, this original example, now in the Smithsonian Institution Castle has been subjected to some alterations during refurbishing. Originally, the finish would have been varied, not all bright gold. The globular shades would have had smaller bases and been made of a lighter, less densely frosted glass. Because reproductions currently on the market unsatisfactorily imitate different styles of shades, restorationists are much in need of better commercially available reproductions.

Since 1964 the Smithsonian Institution Castle has collected over sixty 19th century gas fixtures for restoration. This represents one of the major collections in the United States today in one building.

Plate **36**

Detail of chandelier in plate 35.

I t is by close observation of details such as branches and gas keys that the makers of fixtures may be identified. As already mentioned, this branch (a detail of the chandelier in the previous plate) identifies the fixture as having been manufactured by Starr, Fellows and Company. It also links at least two chandeliers formerly in Quarters One at Springfield, Massachusetts, Armory and two others once in the George Washington Whittemore House in Cambridge, Massachusetts, to Starr, Fellows and Company as well.

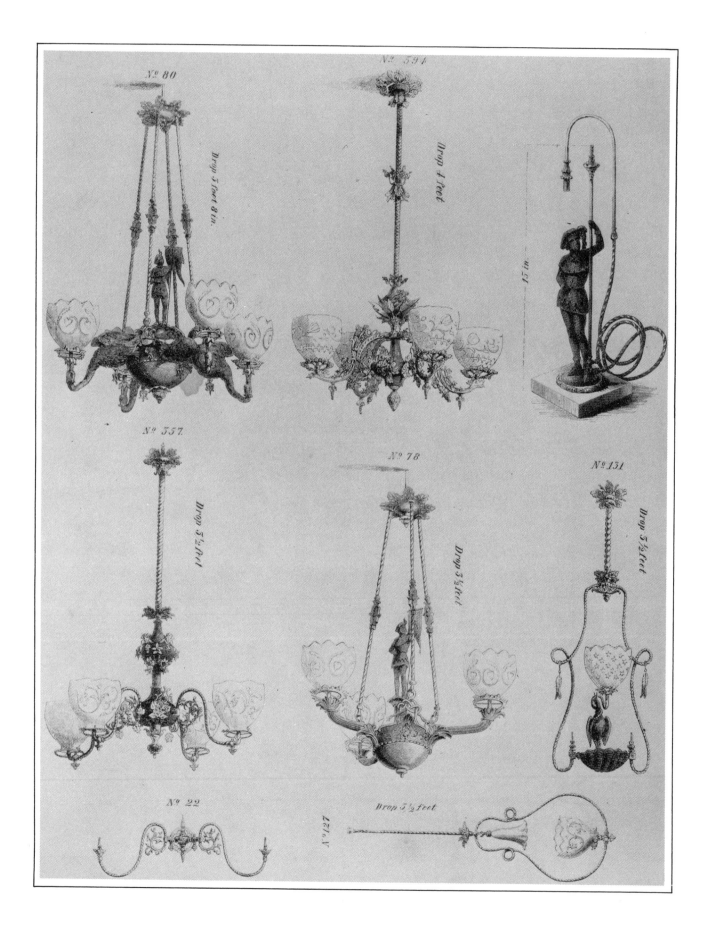

Nº 80

Nº 594

Nº 337

Nº 78

Nº 131

Nº 22

Nº 127

Drop 5 feet 8 in.

Drop 4 feet

15 in.

Drop 5½ feet

Drop 5½ feet

Drop 5½ feet

Drop 5½ feet

This plate, also from the Starr, Fellows and Company catalogue, presents a variety of designs typical of the 1850s. Chandeliers having several rods (no. 78 and no. 80) to conduct the gas instead of a single stem to conduct the gas, were common during the 1850s. Later examples are very rare. Chandeliers no. 357 and no. 594 are of the standard central stemmed type that usually had three, four, or six branches and occasionally five. Bracket no. 22 is of a simple form, and no. 127 (lower right) is a hall light, or pendant, of the characteristic harp type with a glass or porcelain smoke bell. An elaborate variation on the hall harp theme is shown in no. 131.

Rod-hung chandelier, ca. 1856, similar to no. 78 on plate 37.

Plate 38

This chandelier of the rod-hung type has many elements that are similar to, although not identical with, those of the Starr Fellows no. 78 on the previous plate. Because this chandelier has keys precisely like those of the chandelier mentioned on plate 26, it is probable that all are by the same maker. An attribution to Cornelius, Baker and Company seems perfectly plausible in this case; the likelihood is that Starr, Fellows and Company deliberately produced their no. 78 to compete with an already extant design. The finish of this chandelier, one of a pair in the office of the Secretary of the Smithsonian Institution, is dark bronze with gilt accents. The modern shades are incorrect in shape and the quality of the glass, which is too thick and too heavily frosted.

Tudor Place in Washington has a five-rod chandelier that has elements identical in appearance with those of Starr, Fellows and Company's no. 78.[71] The George Washington Whittemore House formerly in Cambridge, Massachusetts, had a three-rod chandelier with many elements, including bowl, branches, keys, and rods, identical with the Smithsonian pair.[72] A three-rod, six-branched example once hung in the library of the T. B. Winchester House at 138 Beacon Street in Boston.[73] As this last chandelier had no elements identical with either Starr, Fellows and Company's chandeliers or those attributed here to Cornelius and Baker, it seems evident that yet another maker manufactured rod-suspension gaseliers.

FELLOWS, HOFFMAN & CO.,

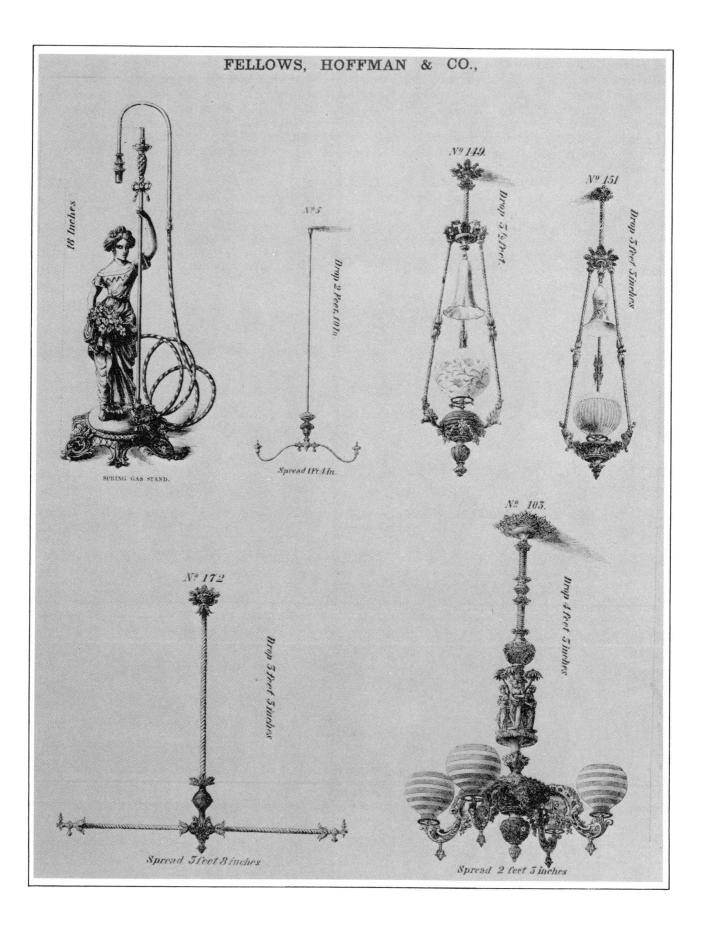

18 inches

SPRING GAS STAND.

Nº 5
Drop 2 Feet 10 In.
Spread 1Ft 4 In.

Nº 149.
Drop 3½ Feet.

Nº 151
Drop 3 Feet 3 inches

Nº 172
Drop 3 feet 3 inches
Spread 3 feet 8 inches

Nº 103.
Drop 4 feet 3 inches
Spread 2 feet 3 inches

Plate **39**

Various fixtures from Fellows,
Hoffman and Company catalogue, 1857-1859.

This plate bears the newly transformed Fellows, Hoffman firm "name and style" and must date between 1857 and 1859. Shown are two hall pendants (top right), a small pendant, a comparatively elaborate "T" pendant (bottom left), a chandelier (bottom right), and a lamp or "stand" (top left). Under the last fixture, the caption "Spring Gas Stand" refers to the allegorical subject of the statuette, not to any mechanism activated by a spring. Another page (not shown) of the catalogue shows a lamp captioned "Franklin Gas Stand" with a Statuette of Benjamin Franklin.

The text of the earlier Starr, Fellows catalogue page headed "Gas Reading Lamps" reads as follows:

> These Lamps or Gas Stands, are furnished with any desired length of tube—6 feet being the quantity usually required, which is prepared exclusively for Gas, and will not leak. The hook is fitted with a universal socket, which will fit any common fishtail or bat-wing burner and is of sufficient length to go over the Glass Shade of the Parlor Chandelier. Elegant Paper Shade Reflectors accompany these Stands.

At least two examples of the chandelier no. 103 are known to exist. One is the collection of Lee B. Anderson in New York City.[74] The other is illustrated by the following plate 40.

Plate **40**

Fellows, Hoffman and Company chandelier illustrated as no. 103 in plate 39, 1857-1859.

If one may judge from the lithographic illustrations of the catalogue, this chandelier, shown in no. 103 on the preceding plate, is as fine in quality and elaborate in design as any made by Fellows, Hoffman and Company. The finish is dark bronze with gilded accents. The forms are decidely Neo-Rococo, with pronounced C-scrolls on the branches. The vogue for figurines clustered about the stem persisted to the end of the 1850s. The shape of the globes is appropriate for the date of the fixture, as well as the etched arcade motif. However, it was unusual to paint the main supporting pipe white or any other color. Normally, it would have been brass pipe lacquered or iron pipe masked by a lacquered brass sleeve. This fine Fellows and Hoffman fixture now hangs in the library of Fountain Elms, an 1850 house museum administered by the Munson-Williams-Proctor Institute in Utica, New York.

GAS PENDANTS.

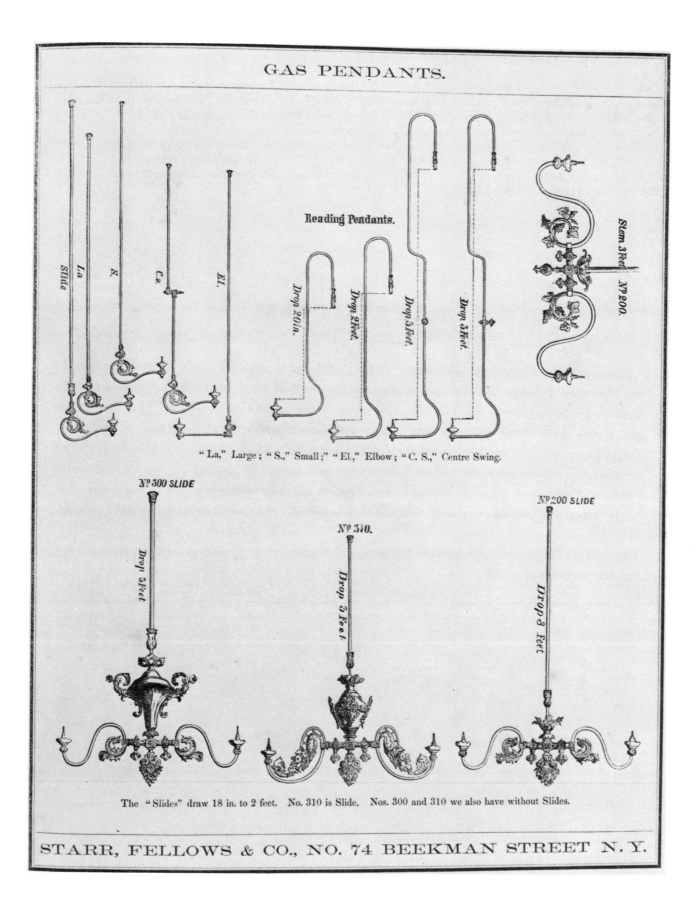

Reading Pendants.

"La," Large; "S.," Small;" "El.," Elbow; "C. S.," Centre Swing.

The "Slides" draw 18 in. to 2 feet. No. 310 is Slide. Nos. 300 and 310 we also have without Slides.

STARR, FELLOWS & CO., NO. 74 BEEKMAN STREET N. Y.

The reading pendants on this plate illustrate that the "stands" attached to a chandelier by flexible tube were by no means the only type of reading light in use during the 1850s. It may be recalled that the lamp referred to in plate 32 was attached by rigid brass tubing like that of which most of the pendants shown here are made. The numbered "slides" were not water-sealed and had no counterweights. Instead, the seal was maintained by the use of cork.

GAS BRACKETS—HALL PENDANTS.

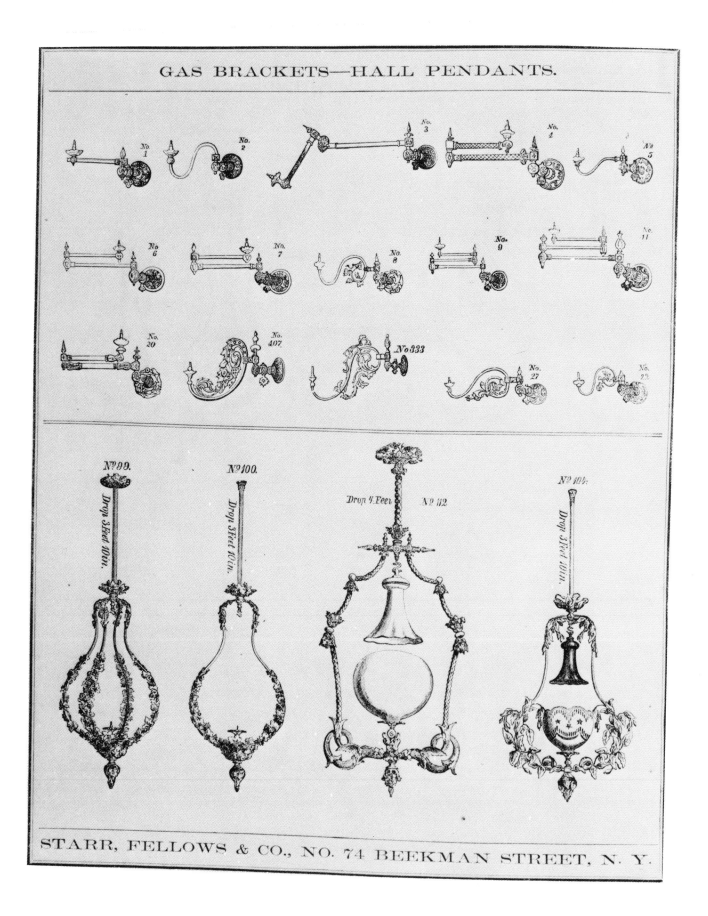

The brackets shown here were designed to allow flexible positioning of the light. All would swing left or right, and several had two or three-sectioned jointed arms. No. 3 (upper middle) had a universal joint at the juncture of the two sections to allow adjustment to any angle. As noted in the discussion of plate 6, jointed branches were often used on bedroom brackets to get light close to mirrors.

The George Washington Whittemore House, formerly in Cambridge, Massachusetts, had a hall pendant of the harp type somewhat like no. 100. The shade of the Whittemore pendant was identical with ones shown on Starr, Fellows and Company bracket no. 40 (not illustrated here). However, attribution by shade design is not advisable as it is probable that the shades were supplied by the glassmakers to various manufacturers of gas fixtures rather than exclusively to one firm.

FELLOWS, HOFFMAN & CO.,

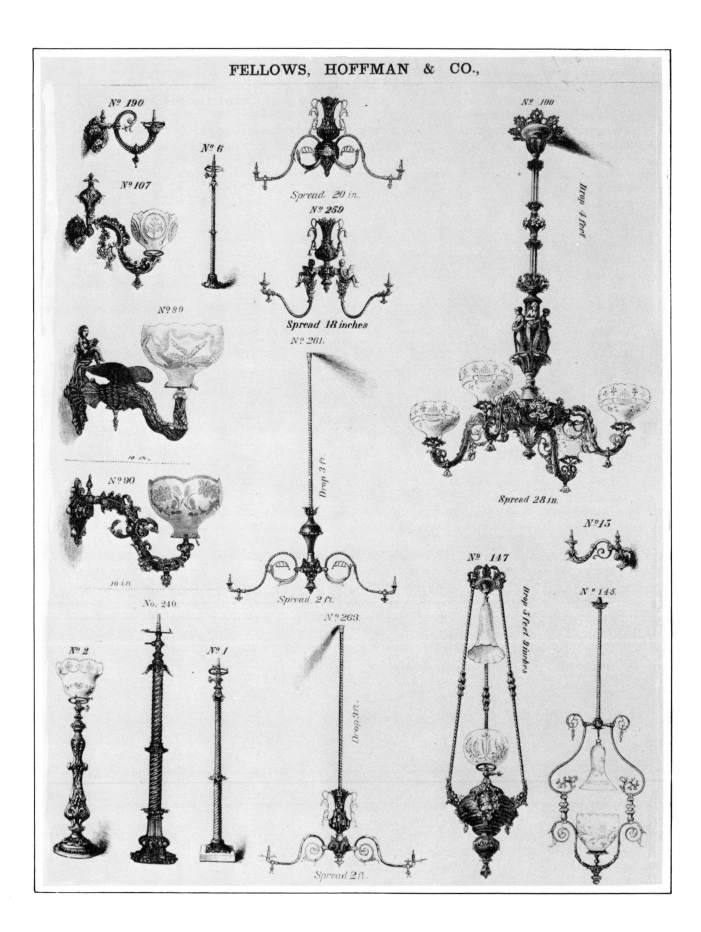

Nº 190

Nº 107

Nº 6

Nº 259
Spread 20 in.

Nº 259

Spread 18 inches

Nº 261.

Drop 3 ft.

Nº 80

Nº 90

10 in.

10 in

Nº 100

Drop 4 feet

Spread 28 in.

Nº 15

No. 240.

Nº 2

Nº 1

Nº 263.

Drop 3 ft.

Spread 2 ft.

Spread 2 ft.

Nº 147

Drop 3 feet 9 inches

Nº 146.

Fixtures nos. 1, 2, 6, and 240 were probably fixed "pillars" rather than lamps or movable "stands." They look too unstable to have served safely as lamps. The seated putto on the birdlike bracket no. 80 (upper left) and his twins on bracket no. 259 (upper middle) would not have had the approval of the critics writing in 1854, who believed that "little figures perched just above the branches . . . violate a fundamental law of decorative art, that all ornamentation should rise out of construction and belong to it."[75] The unnumbered bracket (top center) and nos. 261, 263 (lower center) all share a lily-of-the-valley motif. The hall pendant no. 147 (lower right) and the chandelier no. 100 (upper right) are designated in another place as new designs. The latter is similar to no. 103, shown on plates 39 and 40. Shades of the shape shown on nos. 80, 90, and 100 do not appear after the 1850s. It may also be noted that *no* fixtures in the entire gas section of the catalogue have the decorative chains that were so commonly used during the 1840s.

BRASS FITTINGS FOR BRASS AND IRON PIPE.

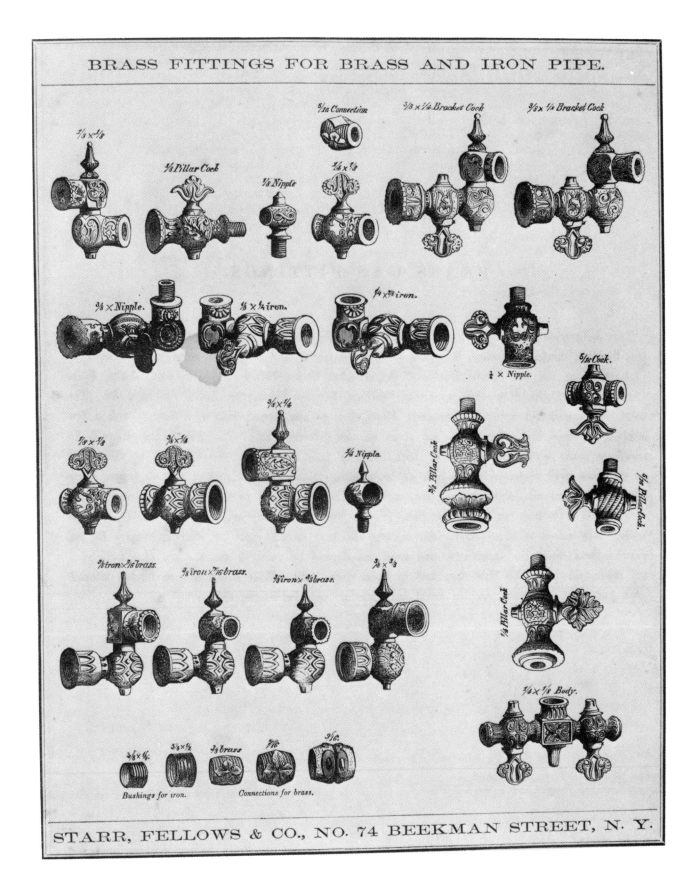

STARR, FELLOWS & CO., NO. 74 BEEKMAN STREET, N. Y.

It is a safe assumption that any unrestored gas fixtures having one or more of the parts shown here was made by Starr, Fellows and Company or Fellows, Hoffman and Company. The use of parts such as these was continued by most companies without change of design (but with the addition of new designs) until the end of the gas era. Although the use by each company of its own designs makes attribution easier, it is not safe to attempt the dating of a given fixture by the presence of a fitting.

This plate of fittings is accompanied in the catalogue by a pedantic text:

> Brass Gas Fittings. This designation is usually understood to include all the *parts* and *pieces*, which go to 'make up' a Gas Fixture; though in reality, it applies only to such *joints* and *connections* as are used to hold the *pieces* together; as Stiff Joints, Double Cocks, Bracket Cocks, Elbow Cocks, Straight Cocks, Pillar Cocks, Swivel Cocks, Top Swings, Centre Swings, Nipples, Connection Balls, Bushings, &c. The word *Fittings* should apply only to such Joints, &c. as are tapped with *iron thread*; and a few samples of these, and other kinds, are given on the following page [i.e., this plate]. The Top Swings, it will be seen, are made with iron thread at both ends, or iron at one, and brass thread at the other, as they are used respectively for iron or brass pipe. *Gas Fittings*, proper, are all tapped with *iron thread*; and all Joints, Swings, &c., with *brass thread*, belong to the *Fixture* department; but it is difficult to keep up the distinction, in the absence of any arbitrary rule to regulate the trade. Nor indeed is it necessary. . . .

BRACKETS

with swing joint. 636.

Another early major manufacturing firm was Archer and Warner, whose history is far more complex than that of the Starr, Fellows—Fellows, Hoffman firm. Founded in Philadelphia, the company had opened a New York branch by 1854. The branch's subsequent history diverged from that of the parent partnership on November 27, 1856, when the original partnership was joined by William F. Miskey to form Archer, Warner, Miskey and Company, a name and style retained until February 27, 1859. In that year, Ellis S. Archer left the company and went to New York. The Philadelphia firm became Warner, Miskey and Merrill, with Redwood F. Warner, William F. Miskey, and William O. B. Merrill as the partners. In 1866 Warner left the firm and Benjamin Thackera joined it to form Miskey, Merrill, and Thackera, a partnership that lasted until 1871. In 1872 Miskey and Merrill were no longer listed in the business section of the Philadelphia directories, and Benjamin Thackera was head of a new partnership. Some of Thackera's work will be discussed later.[76] In the New York City 1854 directory the firm was listed as Archer, Warner and Company. In 1858 the firm name was the same, but James B. Peck was listed as a partner. From 1859 until 1863 the firm was Warner, Peck and Company with Miskey, and Merrill as partners. From 1863 through part of 1866, however, the firm operated under the name and style of Warner, Miskey and Merrill; by 1867 no New York listing appeared.

In the meantime, Archer had left Philadelphia and had formed in New York in 1859 or 1860 his own firm: Archer, Pancoast and Company. The original partners were Ellis S. Archer, George Pancoast, Norman L. Archer, and Anson Archer. In 1861 the partners were listed as Ellis S. and Norman L. Archer, Pancoast (New York), and William C. Ellison (Philadelphia) with Joseph J. Hull as a special partner of limited term. Archer, Pancoast and Company was styled the Archer and Pancoast Manufacturing Company in 1870 and was listed until 1901.[77] Between 1857 and 1859 Archer, Warner, Miskey, and Company commissioned the Philadelphia lithographer Peter S. Duval to execute over 40 handsome color plates illustrating their wares. Although the title of the resulting catalogue is *Warner, Miskey and Merrill Patterns*, the ornamental border of each plate carries the older firm name of "Archer, Warner, Miskey and Company."

All of the fixtures illustrated here (in plate 2 of *Patterns*) are elaborate, and all except the Gothic bracket are mid-century versions of the Rococo style. The colors represented are bronze (light and dark brownish-green in the original tint) and gilt (yellow and rust). Unfortunately, the prices pencilled throughout the catalogue no longer apply. At least one example, no. 698, may have been in stock for a number of years. As shown in the following plate, the pattern of its Gothic branches may have been designed as early as 1845.

Archer and Warner Gothic chandelier.

Plate **46**

This Gothic chandelier was originally in the John J. Brown House, built in 1845 in Portland, Maine.[78] Its branches are identical with those of the bracket no. 698 on the preceding plate, confirming the attribution of this chandelier, now in The Metropolitan Museum of Art, to Archer and Warner. Very few Gothic fixtures of the 1840s and 1850s appear to have been made in America. This surviving example is indeed a rarity. Note that the electrification of this fixture has been unfortunately accomplished by running the wire outside the branches instead of through them. It is admittedly more difficult to run wire through the rather constricted interior of a cast two-mold branch, but the result is worth the effort. Wire of small diameter can be chased through the curving and constricted path by the use of ball chain. Only an experienced electrician of demonstrated ability should be entrusted with this skilled task.

ARCHER WARNER, MISKEY & Co

PL.4

BRACKETS

Brackets from the Archer, Warner, Miskey and Company catalogue, 1857-1859.

Plate **47**

The colors of this lithograph of three Archer, Warner and Miskey brackets are two shades of brownish-green with highlights of yellow with shadings in brown. Clearly, these represent bronze finish with gilt accents. The brackets range from the lavish grape pattern one at the top, priced at $32, to the modestly scaled and comparatively simple example at the bottom, priced at $12 (cf., plates nos. 21, 22 and 23 of this report).

Courtesy of The Metropolitan Museum of Art, The Elisha Whittlesey Fund, 1958.

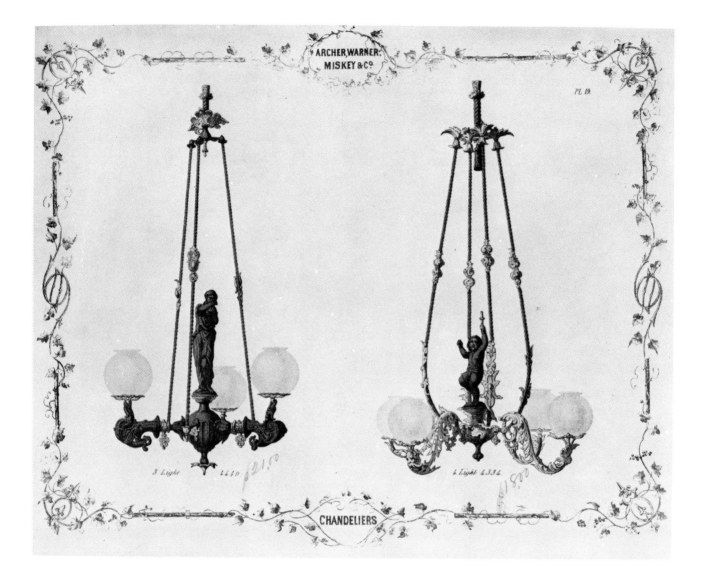

ARCHER, WARNER,
MISKEY & Cº.

3 Light   4440. $215.⁰⁰

4 Light 4334. $180.⁰⁰

CHANDELIERS

Archer, Warner, Miskey and Company was, like Starr, Fellows and Company and other firms, engaged in producing fixtures of the rod-suspended type so popular during the 1850s, as shown here and in the catalogue page shown in plate 23. The rod-supported chandeliers of this period seem invariably to have had figurines above their bowls.

Curiously, the three-light rod-suspended chandelier at the left (no. 4440) was more expensive ($25) than the four-light one (no. 4334, $18). Perhaps the difference represented the higher value of the heavy bronze branches of no. 4440 in comparison with the more ornate gilt branches on no. 4334, possibly made of cast-iron or perhaps white metal. Note that the putto on the right hand fixture holds an extra burner. The metal tassel high above his head adds verisimilitude to the rope motif of the stem and rods.

ARCHER WARNER,
MISKEY & Cº

PL.32.

6 Light 4390

CHANDELIER

Printed in Colors by P. S. Duval & Son, Phila.

Plate **49**

Elaborate chandelier from the Archer, Warner, Miskey and Company catalogue, 1857-1859.

The style of this chandelier and of the previous rod-supported fixtures represent an eclectic blend of Neo-Renaissance, Neo-Baroque, and Neo-Rococo elements. This six-light example cost the then not inconsiderable sum of $75 and is as elaborate in design, though not as large, as any of the 1850s chandeliers made for domestic use. The finish was green-bronze and gilt.
   In quality of materials and workmanship, Archer and Warner and their rivals Cornelius and Baker led the field.

> . . . in 1848 Mr. Archer became associated in partnership with Mr. Redwood F. Warner. Determining to place their house on a level at least with the best in the United States, they saw that this could be effected by merit alone; hence, their first effort was to present in their department of Art, novelty of design combined with superiority of finish and excellence of materials. Sensibly foreseeing that the growing taste in this country required to be fed, they obtained designers and modelers of the highest talents, to whom they paid liberal salaries, and encouraged them in every way to produce graceful and effective designs, for Lamps, Chandeliers, and Gas-fittings. No amount of money was considered by them extravagant, if it secured a valuable result. The consequence of this judiciously liberal expenditure soon became manifest. From an ordinary firm, with a limited capital, doing a moderate business, they sprang to a strong position among the first houses, in their trade, in the United States. Their work is admitted by all to be equal to that of any competitor[79] . . . the manufacture of *Bronzes* . . . I find has assumed, in Philadelphia, more imposing proportions, and a higher artistic character, than it possesses anywhere else in the country. Everybody . . . knows, of course, that the chandeliers, argands, and general gas-fittings of Cornelius & Baker and Archer & Warner are the only American articles of the kind which can sustain a comparison with the goods imported from Paris.[80]

The standing of the firm founded by Ellis S. Archer is attested by the fact that, although Cornelius and Baker were commissioned to light the new Senate and House wings of the U.S. Capitol, Archer, Warner and Miskey received the order to produce the 160-foot-long railings of the two Senate and two House private staircases. These bronze 3-foot-high railings are composed of lavish rinceaux, entwining figures of putti, deer, eagles, pigeons, and serpents executed in full relief from designs by Constantino Brumidi (1805-1880), the Capitol muralist.[81] The brass hemispheres capping the newel posts of three of the four stairways are inscribed "Warner, Miskey and Merrill/ Baudin, Artist/ Phil. 1859." The fourth is date 1858. Edmond Baudin was obviously a skillful sculptor, for whose services "no amount of money was considered by them extravagant. . . ."

Courtesy of The Metropolitan Museum of Art, The Elisha Whittlesey Fund, 1958.

107

PILLAR LIGHTS

"Pillar Lights" from the Archer, Warner, Miskey and Company catalogue, 1857-1859.

Plate 50

This plate of "Pillar Lights" (plate 20 of *Patterns*) shows six fixed standards and four in the form of figurines, that could be used either as fixed or as movable lights. Noted under each of the latter four are both a "Pillar" number and price and a "Flexible" number and price. Rubber hoses supplied lamps as early as 1844 as attested by Thomas Webster's *Encyclopedia of Domestic Economy:*

> Gas lamps may, to a certain extent, be made portable, by having a flexible tube of caoutchouc [rubber] coming from the service gas pipe, and reaching to the place where the gas is required to burn, where it may supply a stand like that of an ordinary candlestick or lamp. This stand may be detached when required, by having one cock at the service pipe, and another at the stand. These are found useful for the desk in offices or other places lighted with gas.[82]

The plate is interesting not only for the variety of designs it displays but also because it shows that as late as 1857 the nomenclature of gas fixtures had not been settled. Whereas Starr, Fellows and Company used the term "lamps" for a few of their stands, the word "lamp" does not appear among these "pillars." Obviously, Archer, Warner, and Miskey regarded the word "lamp" as signifying oil lamp and tried to get around the problem presented by gaslamps through use of the term "flexible."

These fixtures were all finished in greenish-bronze with accents of gilt.

Courtesy of The Metropolitan Museum of Art, The Elisha Whittlesey Fund, 1958.

109

This photograph, taken by C. H. Currier around 1900, shows an interior of a house near Boston with lamp and chandelier probably dating from the mid-1860s. The chandelier has an extra cock to which the connecting hose, or "tube," of the lamp is fastened, thus leaving all three chandelier burners free for use. The visible portion of the inscription on the base of the lamp, or "flexible," is "Patriot," and perhaps the preceding word is "Young." The figure of the boy color-bearer may represent the possibly apocryphal Union drummer-boy who seized his regiment's fallen flag from the hands of a dying color-bearer and bore it bravely forward against the Rebel foe.

All of the burners in the photograph appear to be of the governor type used less frequently in America than in Great Britain.[83] The particular brand seen here was called the "Empire" burner. Governor burners were designed to assure an even flow of gas to the individual jet, but they were, if one may judge from contemporary visual evidence and extant fixtures, not at all common in the United States. Compared to ordinary fishtail or batswing burners, they were quite expensive. Both the lamp and the chandelier have been refitted with late 19th-century wire shade holders and opal glass shades.

An attractive painting titled "Visiting Grandma" by J.A.S. Oertel is dated 1865 and shows a gaslamp attached to a chandelier in precisely the same way as the one in this photograph.[84] The lamp in the painting has a figure of the god Mercury upholding what appears to be a hexagonal shade composed of lithopane panels set in a metal frame.

The quest for the perfect extension gas fixture continued, suggesting that the water seal counterweighted fixtures and the simpler cork seal slide fixtures had certain defects. A major defect, of course, was leakage. Monson's patent, for which this advertisement appeared in 1860, was intended as an improvement over the rather clumsy device of counter-weights.[85] Note that the text stresses safety and durability. Evidently Monson's patent fixtures were used in the buildings erected by the U.S. Treasury Department around 1860, because Major Bowman's endorsement states that they were not merely recommended but actually adopted.[86]

A later publication records the success of such devices:

> Gas escapes frequently occur with water-slide chandeliers, when the water which seals the joints evaporates. The leakage of gas can be avoided either by frequent additions of water, or by putting on the water some sweet oil of glycerine, which retards the evaporation. While water-joint pendants are quite common in England and on the Continent, they are not much used in our country, where either cork-slide pendants or telescopic extension-joint chandeliers are preferred, which dispense entirely with the chain and counterbalance weight and the water seal.[87]

In 1859, Cornelius and Baker supplied the gas fixtures for the rebuilt Vermont Capitol at Montpelier through their Boston agent N. W. Turner at a cost of $2,166. (A pair of 3-foot-high bronze figures of an Indian and a Hunter, cast from the same molds as those supporting the clock in the new House of Representatives in Washington, were also supplied to Vermont at a cost of $75 each.[88]) Fortunately, many of the Vermont Statehouse fixtures remain *in situ*, including the large gilt-bronze and bronze chandelier and brackets seen in this 1860 photograph of the House of Representatives. The statuettes represent, among others, Washington, Franklin, Columbia, Trappers, and Ethan Allen's Green Mountain Boys. The brackets below the gallery balustrade each have a figurine about 10 inches high. One, in compliment to the Vermont-born sculptor Hiram Powers, is a small version of his famous Greek Slave.[89]

By 1859, Cornelius and Baker had supplied the lighting fixtures for nearly all of the state capitols in the United States. The chandeliers and brackets of the Ohio Capitol at Columbus contained, among other embellishments, statuettes of Prudence, Science, Commerce, Liberty, America, and the state's first white settler, Simon Kenton. An American eagle with stars suspended from his beak formed a conspicuous part of the ornamentation. In Nashville, the Tennessee Capitol had in its Hall of Representatives a Cornelius and Baker chandelier 15 feet in diameter decorated with buffaloes, Indians, corn, cotton, and tobacco plants. Clearly, allegory and symbolism could hardly have been carried further in the design of chandeliers.[90] The prestige of such commissions, including the lighting of the U.S. Capitol itself, shows the standing of the firm just before the outbreak of the Civil War. It was then exporting fixtures to India, China, Cuba, South America, and Upper and Lower Canada.[91]

Plate **54**

Cornelius and Baker chandelier in the U.S. Capitol, President's Room, ca. 1860.

Cornelius and Baker's apparatus for lighting the new House of Representatives in the U.S. Capitol was ready by February 2, 1858. The firm provided curving pipes with jets spaced so closely that all could be ignited from a small pilot light. That maze of piping was suspended above the inner stained glass skylight that formed the ceiling of the House.[92] A similar arrangement of 2,500 burners, set so that they could be "lighted instantaneously," was installed above the Senate chamber.[93]

The lighting of other areas of the Capitol in Washington was more conventional. The chandelier shown here just after its electrification in 1896 hangs in the President's Room in the Senate wing. Cornelius and Baker were paid $900 for this 18-branched chandelier on April 9, 1864. It has an astonishing population of bronze figures arranged in three tiers. Their costumes range from almost total nudity, through 17th-century armor, to 18th-century backwoods costume and formal dress on the lower tier. The middle group is composed of putti dancing and playing musical instruments, and the larger-scaled figures of the upper tier include a pioneer or hunter and an Indian.

It would be reasonable to feel that this splendidly rich fixture needed no further elaboration. However, in 1915 the lily was gilded by the addition of six more arms and innumerable *prisms*. This "improvement" brought the total number of lights to an unnecessary 33, 24 of which were visible. The other lights were indirect and concealed in the bowl. That work of supererogation cost $845 in 1915.[94] The shades now on the President's Room chandelier are of a pattern typical of the 1880s.

From the Library of Congress, collection of the Architect of the Capitol.

117

Plate **55**

Cornelius and Baker Indian warrior chandelier
by J. G. Bruff, the Treasury Building, 1859.

Not all fixtures manufactured by Cornelius and Baker were designed by employees of the firm. In 1859 Joseph Goldsborough Bruff (1804-1889), who was then classified as an "ornamental draughtsman," designed a series of chandeliers and brackets under Major A. H. Bowman's direction for the south wing of the U.S. Treasury Building. Bruff left his employment as draftsman in Washington in 1849, recruited 64 men, and formed the Washington City and California Mining Company. He led the expedition overland to California in less time than any other band of 49ers.[95] His gold-seeking, however, was unprofitable.

The details of this chandelier which include Indians armed with bows and arrows and stone tomahawks, wolves, ill-intentioned rattlesnakes, and corn, suggest that Bruff had not forgotten his westward trek. He used these motifs to design an all-American fixture. Cornelius and Baker quoted a price for its manufacture of $50. This quite lavish chandelier was intended for the office of an official of high rank in the Treasury Department.[96]

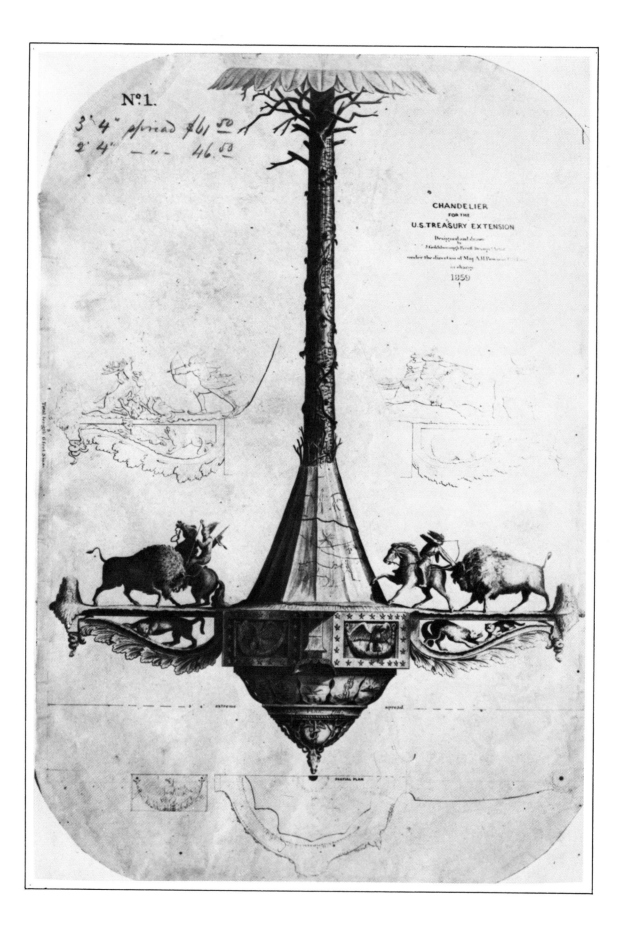

Plate **56**

Cornelius and Baker Indian hunter chandelier
by J. G. Bruff, the Treasury Building, 1859.

This chandelier, the most remarkable of all of J. Goldsborough Bruff's designs for the Treasury's south wing, epitomizes the mid-Victorian concept of appropriate allegorical symbolism. The bowl is inscribed with the arms of each state, and the four-sided section at the springing point of the branches is handled in a classical manner, with eagles perched on garlands within star-bordered panels. In the upper zone, an Indian tepee inscribed with pictographs forms the transition from base to stem, and the stem itself is in the form of a dead pine or redwood. This naturalistic treatment forms the setting for the lively sculptural vignettes of Indian life surmounting the branches. Each of the four branches was planned to carry a different group. At the left a mounted Indian spears a buffalo above the lurking form of a cougar. The lightly sketched branch above indicates an Indian on snowshoes drawing his bow as he confronts a stag attacked by dogs, wolves, or coyotes. At the right a mounted Indian shoots a buffalo with bow and arrow, and underneath are wolves. The sketch above the branch shows a mounted Indian shooting a stag with a rifle, and there appears to be a bear below the scene. All four branches have burners concealed in the forms of tree stumps, below which are Bruff's ubiquitous snakes.

This extraordinary chandelier ornamented with western American scenes was designed for two spread sizes: one 3 feet 4 inches, and the other 2 feet 4 inches. The fact that the larger size cost $61.50 and the smaller $46.50 shows that materials counted for more than labor, because the work involved in making the molds and casting from them was much the same, whatever the size. Bruff designed at least one other chandelier using Indian motifs, a fairly small three-branched fixture 28 inches tall with a 22-inch spread. The overall treatment was conventionally classical, all the foliate motifs being based on acanthus, anthemion, and waterleaf precedents. However, atop each branch was a seated Indian facing the tree stump-concealed burner; one had a bow, one had a rifle, while another smoked a calumet. That rather small-scaled fixture, no. 2 in the series of Bruff drawings, was priced at $16.50.

Nº 6.
$12.

GAS-BRACKET.
Designed and drawn
FOR THE U.S. TREAS'Y EXTEN.
by
J. Goldsborough Bruff, Design'g Artist
Office Treas'y Ex'n Maj. A.H. Bowman U.S.E.
in charge.
1859

Scale of 10 inches.

Cornelius and Baker eagle bracket
by J. G. Bruff, the Treasury Building, 1859.

Plate **57**

Bruff designed a series of ten or more brackets to accord with his elaborate chandeliers. He obviously took pride in his work, because at some point in 1859 he managed to alter his title, or have it altered, from "Ornamental Draughtsman" to "Designing Artist." The bracket shown here treats the eagle and olive branch naturalistically, whereas the rest of the foliage is designed in a conventionally classical manner. The shell seems a bit incongruous, but the gas key in the form of a door key makes a visual pun. Bruff evidently had a sense of humor as well as a zealously whimsical taste.

Two other brackets in the series approached surrealistic fantasies: a woman's arm clothed in an elaborate sleeve and holding a Roman lamp, and a decidedly disagreeable bracket in the form of a rattlesnake. (Perhaps the female arm was a reference to the "Lady with the Lamp" at the hospital in Scutari, the "Florence Nightingale" bracket!) Two brackets had Indian Motifs, one a mounted Indian hunting a buffalo with bow and arrow, the other a standing Indian wearing a bear claw necklace and armed with both bow and arrows and a large mace. He rested one foot on a buffalo skull, while behind him a rattlesnake was about to strike. Three other brackets were more conventional. One recalled the neoclassical, combining a Greek anthemion motif with wheat, acorns and oak leaves, a cornucopia of fruit, and a scallop shell containing a five-pointed star. That last motif appears also in the plaster cornices of the south wing corridors. Another bracket combined acanthus foliage with wheat and corn, and another combined oak, pine, wheat, corn, cotton, tobacco, and grapes, an all-American selection. The gas key of that bracket was in the form of the key on the Treasury Seal.

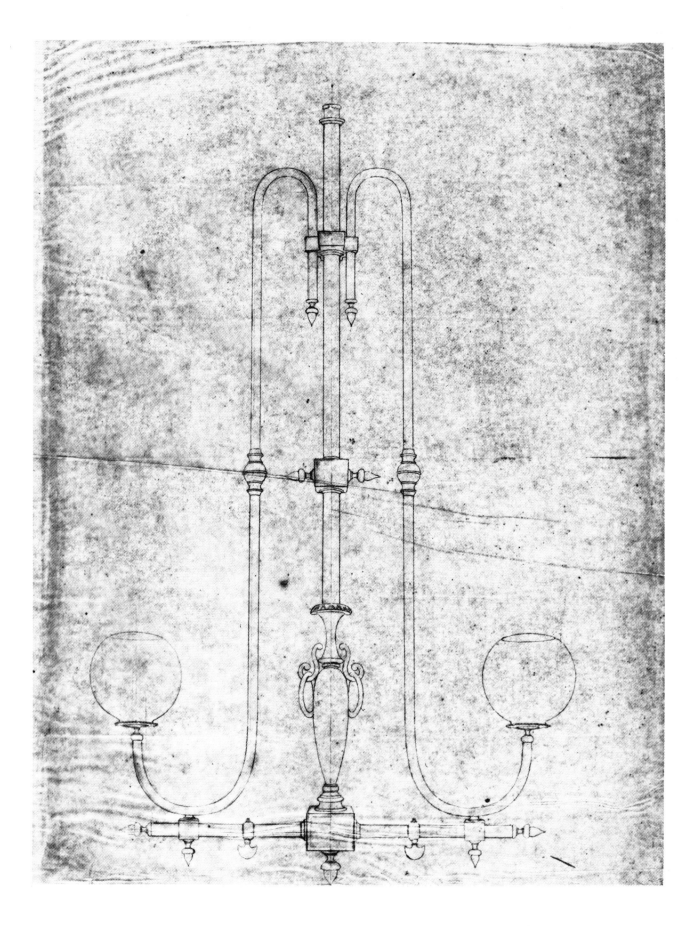

Plate **58**

Cornelius and Baker, simple pendant by J. G. Bruff, the Treasury Building, 1859.

In addition to the lavishly ornamented bronze fixtures just discussed, Bruff drew a series of scaled drawings on tracing paper. These simple fixtures apparently were to be fabricated in lacquered brass. The elaborate fixture shown in plate 55 was also drawn on tracing paper and was numbered 25, evidently the culmination of the series.

The fixture shown here had a spread of 2 feet and 3 inches and was about 4 feet high. It had only two burners and would therefore have been called a pendant rather than a chandelier. It represents a radical departure in style from the narrative or symbolic chandeliers and brackets mentioned earlier. That departure is not merely a matter of simplicity versus elaboration. Instead it lies in a basic approach to design and prefigures the aesthetic revolt that occurred during the 1860s against the ornate Neo-Rococo style, and the often farfetched allegories in miniature sculpture of the preceding decade. That earlier trend had been carried as far as it could go by about 1860, and a reaction was all but inevitable.

Note that all naturalistic forms have been rigidly excluded in the treatment of this pendant. Even the finials suggesting acorns are stylized, and the vase form at the base of the stem has been elongated in a mannerist way. It is not a literal copy of an ancient form, although it is certainly Neo-Grec in character, a style that became the prevailing mode by the later 1860s.

Scale 3 in to the foot

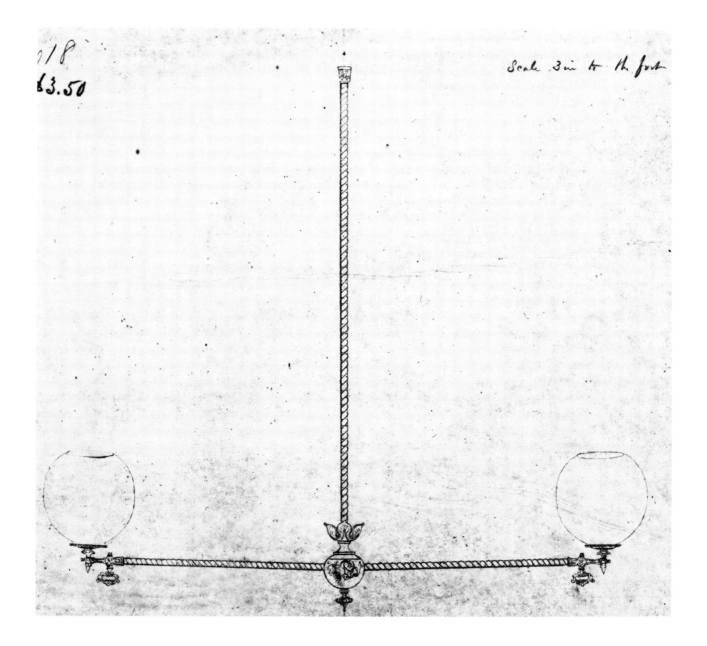

Cornelius and Baker, brass "T"
by J. G. Bruff, the Treasury Building, 1859.

Plate **59**

T his gas "T" of brass rope-twist tubing honors the precepts of the reformist critics of the mid-19th century, such as Horatio Greenough, who held that ornament must arise logically out of construction. There are touches of ornament only at points of junction, the collar at the top, the intersection of the branches, and the keys and burners. The spread of this "T" was a trifle over 3 feet and its height about the same. Presumably the two burners of this $3.50 fixture gave as much light as those of the more elegant $18 pendant shown on plate 58. Clearly, the "T" was intended for a strictly utilitarian area of the Treasury Building.

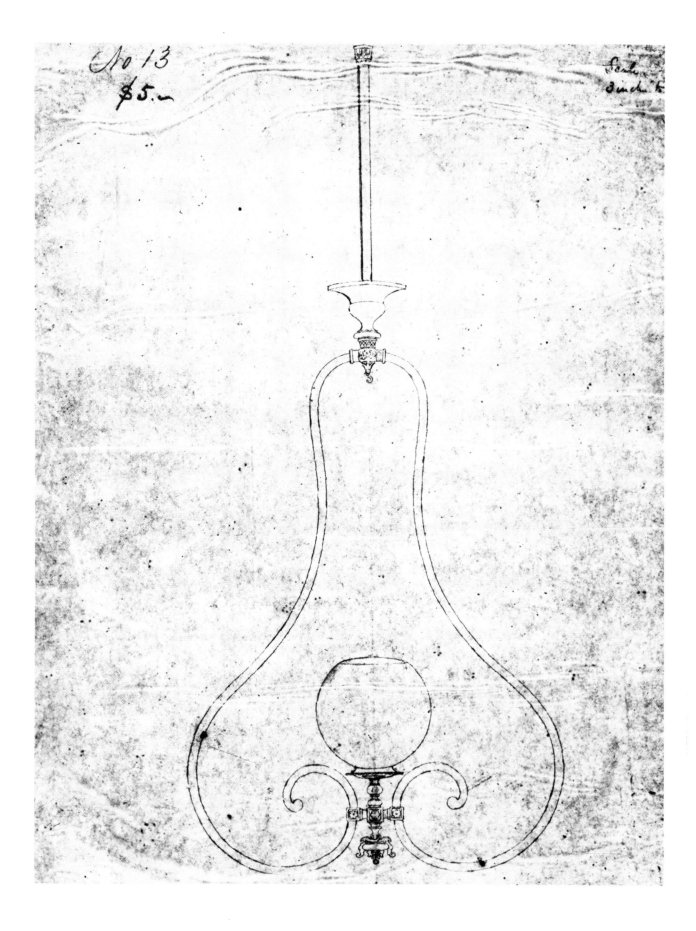

Plate 60

Cornelius and Baker, harp shaped hall pendant by J. G. Bruff, the Treasury Building, 1859.

This hall pendant measured approximately 2 feet wide by 4 feet high and cost $5. Among the Bruff scaled drawings in the National Archives there is one for a slightly larger and rather ornate hall pendant that was to cost $10. The design of the more expensive example was somewhat similar in concept to Fellows and Hoffman's no. 145 reproduced on plate 43 of this report.

The key of the simple pendant shown here, and those of the "T" on plate 59, were designed to have the maker's label in the plain center space: "Cornelius and Baker" on the obverse, and "Philadelphia" on the reverse. Hall pendants of this form were called "lyres," or "harps," or sometimes "lyras" in the last century. The use of so-called "hall pendants" was by no means confined to halls, however, as the example of James Beck's dry goods store in New York has already demonstrated (see plate 30).

Plate 61

Painting of a New York City saloon interior, 1863.

This painting by E. D. Hawthorne in the New York Historical Society shows the interior of George Hayward's Porter House at 187 Sixth Avenue in 1863.[97] Carefully detailed genre paintings such as this are invaluable sources of documentation for their periods. Once the eye has taken in the picturesque variety of New York officers' uniforms, including the Highlander (79th New York Volunteers) at the left and the Zouave in the foreground, other fascinating details become apparent. Among the details are the beer pump at the left, the fly screen at the right, the parquet flooring of alternating light and dark-stained wood, and, certainly not least, the lighting fixtures.[98]

The gilded chandelier and pendants were furnished with smoke bells of porcelain or glass, probably the former since glass is more vulnerable to heat. The chains ornamenting the chandelier were already out of fashion by 1863. Note the manner in which the smoke bells are suspended over each burner. With the exception of lyres, which were almost invariably supplied with a smoke bell because the stem was directly over the burner, smoke bells were not very often used in domestic interiors. However, they were frequently used in saloons, probably to protect frescoed ceilings not only from gas fumes but also from the heavy cigar smoke that was drawn upwards by the convection of lighted gas jets.[99] Though the pendants shown here appear at first glance to be lyres, it is an illusion resulting from the perspective rendering. Actually, they are a most unusual type — two-light pendants with lyraform centers.

W hen the Morse-Libby House (now called Victoria Mansion) was completed in Portland, Maine, in 1863 for Ruggles Sylvester Morse, it was supplied with chandeliers and brackets as splendid as any in America. Unfortunately, no maker's mark appears on them. The 12-light drawing room chandelier and this six-light example in the music room are magnificently executed in gilded bronze. They are, however, among the last important fixtures made in the lavish Neo-Rococo style so fashionable during the 1850s. With the possible exception of the very grand bronze finished chandeliers of Stanton Hall in Natchez, Mississippi, completed in 1857, no extant gas fixtures in America that were designed for domestic use exceed the Morse-Libby chandeliers in scale or complexity of ornament.[100] The globes seen here are original. Note that their holders admitted very little air, being pierced by only a few small holes.

Spelter chandeliers were made in quantity during the 1860s. They were less expensive than the brass or bronze products, as the material itself was very cheap. The major disadvantage of spelter fixtures was their weight. This three-branched example by Cornelius and Baker weighs 20 pounds as against less than 5 pounds for a brass fixture of comparable size. It has been shortened 6 inches by the removal of one section of lacquered brass sleeve and the spelter collars at each end of that piece. The remaining identical section and its collars may be seen on the stem just below the three-handled top ornament. The branches and the palmette-ornamented elements from which they extend are all cast in one piece. These branches, like those on plates 35, 36, 38, 40 and elsewhere, are of the open trough type, leaving the gas tube visible from above. Three kinds of branches were in general use: the open trough type just mentioned; the two-mold cast type shown on plates 18, 26, 76 and others; and the tubular type, usually entwined with ornament, of which examples may be seen on plates 24, 25 and elsewhere.

The eclectic style of this chandelier was probably classifed in its own day as "Renaissance." Among the cast ornaments are four lion heads, three dog heads, and three gilded brass Greek palmettes. The other brass parts are burnished and lacquered. The turned brass shade holders are inaccurate modern replacements. The originals would have been pierced with air holes. The shades themselves date from the right period, although they are not original to this chandelier. They are acid-etched with the same Neo-Grec griffin pattern that is frosted on the later shades of the chandelier on plate 24. Their necks have a diameter of 2½ inches; they are 6½ inches high; and the diameter of the openings at the top is 5½ inches. Their tops are rimmed with gilt copper.

These details showing the obverse and the reverse of a gas key on the fixture in plate 63 have been enlarged for legibility. The key itself measures only 1⅛ inches across. Many other Cornelius and Baker chandeliers were marked this way as well, and the design of the key remained similar, whether it was of cast brass or spelter. As the firm name remained the same for 18 years, precise dating based on the wording of the mark is not possible.

During most of the 1860s the advertisers taking the best display space in *The American Gas-Light Journal* were Cornelius and Baker and the New York firm of Mitchell, Vance and Company. A Fellows, Hoffman and Company advertisement often occupied the next best spot. This confirms indications that these three were the leading firms of the decade. Evidence to be mentioned later will show that by 1876 Mitchell, Vance and Company had surpassed Cornelius and Baker and achieved the leading position.

In 1868 the Philadelphia firm's partnership was composed of Robert Cornelius, Isaac F. Baker, William C. Baker, Robert Comeley Cornelius, John C. Cornelius, Robert C. Baker, and Charles E. Cornelius. During 1869 the famous partnership formed in 1851 was dissolved, and by 1870 Cornelius and Baker had become Cornelius and Sons.

THE PARIS UNIVERSAL EXHIBITION.

We engrave on this page four CHANDELIERS of Cast

Iron, manufactured and exhibited by Mr. TUCKER, of

New York and Boston. America gives but scant

material for introduction into our Catalogue, but

these productions are of considerable merit; as mere castings they are unsurpassed. The designs also are of more than ordinary value. Their peculiar worth, however, is derived from a new and very interesting process of manufacture, to describe which here is impossible, but to which we shall elsewhere direct public attention.

Plate **65**

Engraving of Tucker Manufacturing Company
cast-iron chandeliers in the Paris Universal Exhibition, 1867.

The only American gas fixtures mentioned in the catalogue of the Paris Universal Exhibition of 1867 published by the *Art-Journal* were those illustrated here by the Tucker Manufacturing Company of New York and Boston.[101] Further significant recognition was accorded the firm (actually headed by Thomas J. Fisher) about a year later when Alfred Bult Mullett, then Supervising Architect of the Treasury Department, commissioned the Tucker Manufacturing Company to make the fixtures for the new north wing of the Treasury Building in Washington. Among the fixtures were those for the lavishly ornamented Cash Room, the setting for President Grant's first inaugural ball. The Treasury boasted, possibly with less than complete justification, that the Cash Room was "the most costly room in the world."[102] (See Preface for correction.)

The rather condescending text of this illustration reads as follows:

> We engrave on this page four Chandeliers of Cast Iron, manufactured and exhibited by Mr. Tucker of New York and Boston. America gives but scant material for introduction into our Catalogue, but these productions are of considerable merit; as mere castings they are unsurpassed. The designs are of more than ordinary value. Their peculiar worth, however, is derived from a new and very interesting process of manufacture, to describe which here is impossible, but to which we shall elsewhere direct public attention.[103]

Cast iron enjoyed so great a vogue during the 1850s and 1860s that it is not astonishing to find it used for such elaborate gaseliers as that on the right. Although its best known use was for architectural elements, including entire facades, fences, garden furniture and the like, cast iron was so popular that it was even used for indoor furniture such as hall stands, small tables, plant stands, and footstools, among other things.

These cast-iron fixtures are in the latest style current at the time of their exhibition. They are austerely stiff and angular compared with their earlier counterparts and presage the Eastlake manner soon to follow. Although Charles Locke Eastlake would not have given full approval to the chandelier at the right, he would probably have found the middle and lower fixtures at the left sufficiently "honest" in their expression of "constructive principles" to merit approbation.[104] Even the most elaborate of these gaseliers with its tight and rigid curves evidences a strong reaction against the lush mode of the preceding decade.

From the author's collection.

When it was completed in 1869, Alexander Turney Stewart's marble mansion on the northwest corner of Fifth Avenue and 34th Street was popularly regarded as the most sumptuous residence in New York City.[105] The hall, reception room, music room, and master chamber had 12-light bronze chandeliers of similar, if not identical, conventionally ponderous design. The drawing room chandeliers shown here, while equally complex in design, are much lighter and more delicate in treatment. The branches have the elaborate but slender character that became so fashionable toward the end of the century. In that respect, they are forward looking. Note the urn-like porcelain elements ornamenting the stems of these fixtures. They portend a more wide-spread use of porcelain on chandeliers during the following two decades.

At first glance, the shades appear to be of the type that came in around 1880, but close scrutiny reveals that they still have the very constricted necks of an earlier date. Note that they are very lightly etched. Their delicate patterning leaves them, for all practical purposes, almost clear.

The standing fixtures, or "lampadaires," flanking the far window were not commonly found except in the most lavish domestic interiors of the period. They solved the problem of achieving extra light without using brackets, which would have interfered with the frescoed design of the wall panels. All the Stewart Mansion fixtures, of course, were exceptional in both scale and grandeur.

The chandeliers in the library of A. T. Stewart's mansion had more angular branches, Egyptian Revival heads, stiff palmettes, slender beaded chains, and elaborate colonnettes —all typical Neo-Grec elements of the Neo-Renaissance style (now frequently called "Centennial Renaissance") that flourished from the mid-1860s until after 1876. Chandeliers with clusters of fancifully formed rods or colonnettes arching out from the central stem and surrounding it were very fashionable during the 1870s.[106] However, the most significant innovation to be seen on these Stewart Mansion fixtures of 1869 is the double cone reflector at the base of each chandelier. The construction and function of mirrored reflectors of this type will be discussed later in plates 86, 87, and 88. They were adjustable, permitting light to be aimed and concentrated at will.

Unlike the library, the famous picture gallery in the Stewart Mansion was provided with gas reflector fixtures alone, without conventional burners.[107]

Plate **68**

Section through a mansion showing mechanism of a Springfield Gas Machine, ca. 1868.

So-called "portable gas works" or "portable gas machines" were in use as early as 1813, when David Melville of Newport, Rhode Island, patented a gas machine in March of that year. In 1846, the Long Island Sound steamboat *Atlantic* had portable gas works aboard.[108] They were used principally to generate gas for consumption where centrally manufactured gas was not available. Occasionally, establishments using large quantities of gas found it economically advantageous to generate their own illuminant even when commercial gas was at hand. That was the case with the 600-room St. Nicholas Hotel of 1854 in New York City, as a contemporary advertisement relates:

> A spacious private gas-house, capable of furnishing 200,000 cubic feet of gas per night, supplies the hotel with the material of light. This building, like the steam-generating department is also detached from the main structure . . .[109]

Another advertisement says:

> The gas light here is made from resin and costs only $924 a month, whereas if the hotel bought it from the gas companies it would cost over $2,500 for the same period yet it does brightly illuminate all the rooms and for those who understand its workings, is no menace at all.[110]

Various substances, including some extremely volatile and dangerous ones, were used to supply gas machines. James J. Walworth's gas generating system for residences, churches and public buildings supplied gas made from resin and was in operation by 1850.[111] O. P. Drake, a Boston manufacturer of chemical and "philosophical" (i.e., scientific) apparatus, advertised "Drake's Patent Hydro-Carbon Gas Generating Apparatus" in 1854. The apparatus was designed for "Private Dwelling Houses, Churches, Hotels and Factories, in the country, where coal gas cannot be obtained without great expense." The distillate to be vaporized was "benzole" (benzine). Drake's prices ranged from $150 for a six-light apparatus to $800 for a 100-light one.[112] There were also devices for enriching coal gas by "carbonizing" or "carbureting" it. John Amsterdam patented a method of carbonizing illuminating gas on June 15, 1858.[113] This print, drawn and engraved by John Keim and published around 1868 by Hay Brothers, shows one of the celebrated Springfield Gas Machines installed. The mansion it serves, a grand example of mansard-roofed splendor in the French Second Empire manner, is evidently imaginary, since, except for the hall, its elegant interiors are all drawing rooms. (There is not a dining room, library, or chamber to be seen among the parlors.) Note that the ample grounds of the magnificent imaginary estate contained at least three lamp standards and that the stable was also gas-lighted. This was probably more advertising hyperbole than truth.

The Springfield Gas Machine appears to have been one of the most successful contrivances of its kind in use during the last third of the 19th century. It generated gas from gasoline, a fuel supplied by the company as early as 1865.[114] The explosive danger of vaporized gasoline was recognized by placing the generator, or evaporating tank, underground at a distance from the building or buildings to be lighted. Presumably the risk to the person whose duty it was to replenish the tank was accepted as the necessary price of "Progress," much as steam boiler explosions seem then to have been regarded. (Lest the last century be thought more callous than our own, it should be remembered that the carnage caused by automobile accidents now seems acceptable.) An air pump driven by a weight suspended from the cellar ceiling forced air through a pipe leading to the generator, and a second pipe conducted the gas from the underground tank to the house.[115] The apparent prevalence of "portable gas machines" clearly indicates that the absence of centrally manufactured gas in a community by no means precludes the possibility that structures in that community were, in fact, gaslighted.

Painting of W. H. Vanderbilt Parlor, New York City, 1873.

Plate **69**

Visual documentation of American gaslighted interiors is relatively scarce. One of the best post Civil War genre paintings showing the effect of gaslight is this example painted in 1873 by Joseph Seymour Guy. It represents William Henry Vanderbilt and his family in the cozy back parlor of their bourgeois brownstone before their famous Fifth Avenue mansions and Newport "cottages" were planned. The room is lighted by a pair of angular brackets and a chandelier. The central light could be lowered for reading. A large and splendid unlighted glass chandelier is in the front parlor beyond the arched doorway. Note that the bronze candelabra of the mantel garniture are not being used. The library of Alfrederick Smith Hatch's house at Park Avenue and 37th Street in New York, shown in Eastman Johnson's 1871 painting of the Hatch Family, had a chandelier very similar to this W. H. Vanderbilt back parlor fixture.[116]

Courtesy of the Biltmore Estate, Asheville, North Carolina.

Plate 70

Chandeliers installed in the East Room
of the White House, 1873.

In 1873, President Grant had the up-to-date fixtures seen in this photograph replace the three East Room glass chandeliers that had been bought for the White House by President Jackson in 1834 and fitted for gas in 1848 (see plate 13). This J. F. Jarvis stereograph was taken after the room was redecorated for President Arthur in 1882-1883 by Associated Artists. These immense chandeliers, hung with dozens of notched spear prisms and lighted by a multitude of shaded burners and internal reflectors as well, were as splendid as any in America of their day. Such chandeliers represented the ultimate in grandeur and luxury to most men and women of the gilded age. A contemporary publication, describing a Fall River (Massachusetts) Line steamboat, spoke of "costly chandeliers in the sunbeams darting forth the bright rays of the prism, or by gaslight sparkling with all the brilliancy of a tiara of diamonds."[117]

From the author's collection.

Detail of a glass chandelier showing ball and socket joint at ceiling, ca. 1865.

Plate **71**

Although glass, or "crystal" chandeliers were by no means unknown, particularly in the Boston area before the Civil War, the real vogue for them came in the last third of the 19th century. During the 1850s and 1860s frosted and cut glass bowls and stem balusters and frosted bobeches were often used in combination with clear notched spear prisms. Frosting the glass was a means of minimizing the effect of the silvered pipes to which the clear glass branches were attached. The glass branches were fragile and could not withstand very much torque. It was therefore necessary to hang glass chandeliers so that they were not rigid. "Larger and heavier fixtures are hung with universal ball-and-socket joints."[118]

The chandelier of which a detail is shown here formerly hung in the J. H. Bancroft House in Cambridge, Massachusetts, and probably dated from around 1865. It dates from before 1873 and is almost certainly of Massachusetts origin. It is 33 inches wide and just short of 70 inches tall. The prisms, or "icicles" are 5-1/2 inches long. Note particularly the ball-and-socket joint where the pipe meets the ceiling. The elaborate center flower was not molded plaster but simply illusionistically painted in grisaille on a perfectly flat surface.

A similar but slightly larger chandelier fitted with gas candles is in The Metropolitan Museum of Art and is attributed to Gillinder and Sons of Philadelphia.[119] It is quite unlikely that the Cambridge chandelier came from Philadelphia; however, it is very likely that it was made by an East Cambridge firm. The Cambridge fixture and the supposedly Philadelphia fixture are so similar, it seems probable that the Metropolitan Museum's example should be attributed to a Massachusetts firm.

EXHIB. MT. WASHINGTON GLASS WORKS EXHIBIT. M. 80

Mt. Washington Glassworks Display
in the Centennial Exhibition, Philadelphia, 1876.

Plate 72

The Centennial Exhibition of 1876 in Philadelphia exhibited a wide variety of products. Lighting fixtures were certainly not least among the categories shown. Among the many groups of manufactures exhibiting American prowess were the wares of the Mount Washington Glass Works of New Bedford, Massachusetts, seen here. The dazzling display of crystal chandeliers and glasswares included epergnes and other tablewares, vases, and at least three painted glass chandeliers, one at the extreme left and two, one behind another, left of center. Note also the various sizes of shades stacked within the counter at bottom center of the photograph. This group of fixtures includes most of the styles current in glass chandeliers during the 1870s.[120]

Brass and ormolu chandelier
by Mitchell, Vance and Company, 1870.

Plate 73

This handsome brass and ormolu 12-light chandelier hung in the sitting room of the now-demolished Jedediah Wilcox House completed in 1870 in Meriden, Connecticut. The chandelier (5 feet 8 inches tall) was never electrified. The lower six burners have slotted shafts with slotted sliding brass sleeves to permit varying amounts of air to be mixed with the gas, to adjust the pressure. The classical profile heads in high relief just above the gas keys, the vase and urn motifs of the stem, and the stylized palmettes are all hallmarks of the Neo-Grec style of the late 1860s and the 1870s. The etched and cut glass shades are original.[121]

The Wilcox House chandeliers were made by the New York firm of Mitchell, Vance and Company. Originally, the firm was Mitchell, Bailey and Company, incorporated in 1854 in Connecticut by John S. Mitchell, John Bailey, Anson H. Colt, and Samuel B. H. Vance. In 1860 the new copartnership of Mitchell, Vance and Company was organized by Mitchell, Vance, and Aaron and Charles Benedict. In 1873, the Connecticut copartnership was dissolved, and Mitchell, Vance and Company was reincorporated in New York. When John S. Mitchell died on February 1, 1875, the firm was managed under the same name by Charles Benedict, President. As early as 1856 the firm had a fashionable Broadway address. In 1877, they built offices at 836-838 Broadway, where they were still listed in 1902.[122]

A Mitchell and Vance advertisement in 1881 read: "Mitchell, Vance and Company Manufacturers of Gas Fixtures, Fine Clocks, and Bronzes. Highest Award and Medals at the Centennial Exhibition. Crystal, gilt, bronze, and decorated [i.e., polychromed] gas fixtures in the greatest variety at low prices. Special designs for churches, halls, hotels, dwellings, etc."[123]

Note that the firm received the highest award at the Centennial Exhibition of 1876 in *Philadelphia*. Cornelius's output may have been the largest in volume, but by 1876 Mitchell, Vance and Company led in prestige. The judges' report for the 40th Exhibition of the American Institute said: "The Glass Chandeliers are equal, if not superior to the celebrated Osler manufacture [English], which have been . . . the best in the world. The Glass is of unusual whiteness. They Rank A1."[124]

The firm supplied the fixtures for such prestigious buildings as: St. Patrick's Cathedral, St. Thomas's Church, Collegiate Church of St. Nicholas, and Temple Emanuel, all on New York's Fifth Avenue, and H. H. Richardson's Brattle Square Church in Boston, Boston City Hall, and the new Illinois State Capitol. There are 69 structures listed in Mitchell, Vance and Company's *Centennial Catalogue* including such hotels as the Windsor and the Grand Central in New York, the Palmer House in Chicago, Galt House in Louisville, and the Grand Union and United States Hotels in Saratoga Springs. Theaters included the Booth Theatre and the Grand Opera House in New York City. College buildings listed were Harvard's new Memorial Hall and "Smith's Female College" in Northampton, Massachusetts. Commercial Structures included two of the most important buildings of the time, the Western Union Telegraph Building and the Tribune Building, New York proto-skyscrapers.[125]

The text of the *Centennial Catalogue* indicates contemporary practice. "For the Reception Room, Chandeliers in Gold, relieved with a little color—as jet, crimson, or blue are deemed desirable." A matching 12-light chandelier was suggested for the drawing room, and chandeliers with center slides for "other rooms," presumably the dining room and library. Among these was a seven-light (six-branched) "slide Library Chandelier in the Neo-Grec style." The center slide had an Argand burner and could be lowered "very near the reading table." That fixture was ornamented with medallions representing music, poetry, and history, but the substitution of medallions representing game, birds, and fish could render it suitable for the dining room. It was available in bronze, gilt, or verde antique finish. The "Standards" shown were pillar lights with several burners and included a crystal and an ecclesiastical design. One newel standard was supported by a bronze American Indian. The grandest crystal fixture illustrated was a 30-light chandelier. The most detailed description was reserved for the lavish Neo-Grec eight-light chandelier made for the main entrance of the Western Union Telegraph Building. This fixture had a laurel-garland Greek vase, four fluted colonnettes with foliated capitals, and burners in the form of classical lamps. It also had an extraordinary assortment of fauna—female nudes in low relief, lion heads, griffins, and animals that appeared to be the progeny of sphinxes mated with unicorns. It was designed by Charles C. Perring, whose skill probably contributed much to the outstanding success of the firm.[126]

2295—GAS FIXTURES

Plate 74

Cornelius and Sons display in the
Centennial Exhibition, Philadelphia, 1876.

Tꜱhe Cornelius and Sons' kiosk in the main building at the Centennial Exhibition shows that at least one large glass chandelier was in their repertoire, although their 22-page catalogue of about the same date shows none. Certainly crystal fixtures were not their usual product. Most of their metal fixtures of the 1870s were decidedly angular in design and were derived more from a misinterpretation of Eastlake principles than from a sophisticated understanding of the Neo-Grec style. Many of Cornelius and Sons' designs in their undated catalogue, probably issued in 1876, appear awkward and naive in comparison with those of Mitchell, Vance and Company, although that is by no means invariably the case. Note that the table in the foreground bears several lamps whose burners are supported by bronze figures. Two of those in the catalogue represent seasons and are labeled "Ete" and "Hiver," which suggests the prestige of "French bronze," even when made in America.

The history of the Cornelius firm down to 1870 has already been traced.[127] After the split with the Bakers in 1869, Cornelius and Sons in 1870 was composed of Robert Cornelius and his sons Robert Comeley, John C., and Charles Blakiston, Samuel Loder and Albert G. Hetherington. Another name change occurred in 1886, when Cornelius and Hetherington (John C. Cornelius and Albert G. Hetherington) were the partners. An 1887 advertisement read as follows: "Cornelius and Hetherington — Artistic gas and electric fixtures, wrought iron and brass grills, memorial brass, real bronze, railings and castings. 1332 Chestnut Street."[128] From 1888 until the firm dissolved in 1900 it operated as Cornelius and Rowland with John C. Cornelius and George L. Rowland as partners.

The Baker group formed Baker, Arnold and Company by 1871 or the year before. In 1875 the partners were listed as William C. Baker, Crawford Arnold, and Robert C. Baker at the old Cornelius and Baker address, 710 Chestnut Street. Crawford Arnold first appeared in 1859 at the same address and apparently was a member of the firm of Cornelius and Baker but not a partner. Baker, Arnold and Company was last listed in 1878.[129]

No. 6492. 12 Lights.
Length 53 inches, Spread 31 inches.
No. 6493 GILT.

No. 6492. 1, 2 and 3 Lights.
Spread 17 inches.

No. 6410. 12 Lights.
Length 55 inches, Spread 31½ inches.
No. 6411 GILT.

CORNELIUS & SONS, No. 821 CHERRY ST. PHILADELPHIA.

The undated Cornelius and Sons catalogue previously mentioned is in The Historical Society of Pennsylvania collection in Philadelphia and is the only copy known to exist. It is composed of 22 lithographed plates showing a variety of fixtures, most of which are depicted in a brown color representing bronze, although a few are shown as gilded. The angularity of the three fixtures on this plate is typical of many that were made around the centennial year of 1876. The chandelier at the left has elements of Neo-Grec style, but the chandelier at the right, an eclectic blend, is more Eastlake than otherwise in design. These bronze finished chandeliers, nos. 6492 and 6410, had gilded counterparts, nos. 6493 and 6411. Twelve-light chandeliers were never very common. They were used only in large and grand houses and in public buildings.

Plate **76**

Cornelius and Sons chandelier illustrated in the previous plate.

This is one of three gilded chandeliers catalogued as no. 6411, the gilt version of no. 6410 at the right of the preceding Cornelius and Sons plate. The three measure 53 inches high and have a spread of 31½ inches. Two hang in the drawing room and one in the *en suite* dining room of an Alexandria, Virginia, residence built between 1850 and 1855. Alexandria already had city gas in 1851, so these fixtures of the mid-1870s must be presumed to have replaced earlier chandeliers. As damage to earlier chandeliers was unlikely, they were probably removed by a previous owner. Frequently fine fixtures were considered to be furnishings rather than fittings and were retained by the seller when a house changed hands.

Courtesy of Mr. & Mrs. Bernard Fensterwald, Alexandria, Virginia, photograph by Jack E. Boucher.

161

Detail of Cornelius and Sons chandelier in the previous plate.

Plate 77

This detail of the chandelier in the Eastlake manner by Cornelius and Sons shown on the preceding plate clearly demonstrates that, although the designs may have fallen short of the firm's previous standards, the quality of workmanship was fully maintained. The execution of complicated castings was every bit as well carried out by Cornelius and Sons as by any of their rivals.

Courtesy of Mr. & Mrs. Bernard Fensterwald, Alexandria, Virginia, photograph by Jack E. Boucher.

Plate 318

No 6736, 3, 4 & 6 Lts
Length 48 in   Spread 23 in.

$13.00

$16.00

No. 6670, 2, 3 & 4 Lts.
Length 34 in.
Spread 21 in.

No. 6494, 2, 3, 4 & 6 Lts.
Length 39 in
Spread 23 in.

CORNELIUS & SONS, No. 821 CHERRY St. PHILADELPHIA.

Plate 78

Chandeliers from Cornelius and Sons catalogue, probably 1876.

The notation "plate 318" in the upper right corner of this plate suggests that there was a group of illustrations in print much larger than the 22 lithographed plates that comprise the extant Cornelius and Sons catalogue (see plate 75 of this report). The 821 Cherry Street address refers to the company's factory, not to the retail outlet, which was at 1332 Chestnut Street from 1870 until the late 1890s. The stem of the chandelier in the middle and the griffins perched on the branches of the chandelier at the left show Neo-Grec influence, but the fixture at the right almost defies stylistic analysis. Note that the branches of the chandelier at the left are identical with those of the 12-light chandelier just discussed (except for the griffins, see plate 77). These three fixtures could be had with from two or three to six lights. They range from 21 to 23 inches in spread and from 34 to 48 inches in height. Each shade was secured by a single set-screw. Two short claws (not visible in this lithograph) on the supporting ring engaged the lower lip of the shade and held it firmly.

Cornelius and Sons bracket with griffin, ca. 1876.

Plate 79

The cast griffin ornamenting this Neo-Grec gilded bracket is identical to those shown on the chandelier at the left of the preceding plate. The bracket can therefore be attributed with certainty to Cornelius and Sons. The inappropriately plain spun brass wall plate is a modern replacement for an undoubtedly more ornate lost original. Wall plates were invariably used with brackets to mask the break in the plaster where the gas pipe emerged from the wall. They were often shallow in profile because there was no need, as in modern installations, to accommodate wires and wire nuts. The shadeholder of this bracket is a modern restoration. But the design is reasonably suitable for shades of post-1880 vintage, although three set-screws instead of one, as in original shade holders, are used. The shade dates from the 1880s and is etched with Neo-Grec patterns. Probably the original shade was still of the small-necked variety.

From the author's collection, photograph by Jack E. Boucher.

Plate 320

Nº 6528, 6 & 9 Lts.
Length 26 in.
Spread 19 in.

Nº 6770, 6 & 9 Lts
Length 22 in.
Spread 19 in

Nº 6772, 9 Lts.
7 ft 6 in high, Spread 19 in

Nº 6690, 6 & 9 Lts.
Spread 25 in

CORNELIUS & SONS, No. 821 CHERRY ST. PHILADELPHIA.

Plate **80**

Polychromed Gothic Revival fixtures
from Cornelius and Sons catalogue, probably 1876.

The 7½-foot high standard and brackets shown here in the undated Cornelius and Sons catalogue were meant to be ecclesiastical fixtures, polychromed with accents of red and blue. They were probably designed by J. M. Beesley.[130] A precedent for this polychromed, or "decorated" standard of ivy-ornamented Gothic Revival pattern (and also for the use of large coronas in church lighting) was set at least as early as 1853, by the use of similar gas standards in the restoration of the 15th-century St. Botolph's Church in Boston, Lincolnshire. That major English parish church installation was illustrated and described in the *Illustrated London News* as follows:

> The arrangement of the lights is novel and successful. Instead of the usual plan of solitary brackets scattered ineffectively over the church, there are rich brass standards, each bearing a considerable number of jets, and producing a vista of light. Over the font is suspended a magnificent corona bearing nearly a hundred lights. The adaptation of the modern invention of gas to ancient churches, so as not to destroy the effect of their architectural structure by incongruous fittings, has long been one of the most vexed problems of church restoration. The most fastidious stickler for ancient precedent would acknowledge that the richly-decorated standards and the crown of light at the western end harmonise so entirely with the whole building in its restored aspect, that they might almost be deemed part of the original design.[131]

Coronas such as that installed by Cornelius and Sons in the Columbus Avenue Universalist Church in Boston in 1873 and the one (probably by Mitchell, Vance and Company) that hung in Trinity Church, Boston from 1877 until the 1930s were apparently popular for major American Victorian Gothic churches of the 1860s and 1870s. The original model for them was probably the magnificent corona given in 1168 by Friedrich Barbarossa to Charlemagne's Palatine Chapel at Aachen (Aix-la-Chapelle).[132]

Nº 0108 - 3 Lts.
Projects 10 in.

Nº 0110 - 3 Lts.
Projects 12 in.

Nº 0108½ - 6 Lts.
Spread 15 in.

Nº 0109½ - 2 & 3 Lts.
Spread 14 in.

Nº 0101 - 1 Lt.
Projects 17 in.

Nº 0101½ - 2 & 3 Lts.
Spread 24 in.

Nº 0103½ - 2 & 3 Lts.　　　Spread 19 in.

Archer & Pancoast M'f'g. Co.

MANUFACTORY AND WAREROOMS Nºs 70,72 & 74 WOOSTER ST. N.Y.

Plate 81

Gothic Revival brackets from Archer and Pancoast Company catalogue, probably 1876.

The Archer and Pancoast Manufacturing Company had a prominent display at the Centennial Exposition of 1876 in Philadelphia. This plate shows that the New York firm, founded by Ellis S. Archer, competed with Cornelius and Sons in the church fixture field with their own version of the Gothic Revival ivy pattern, albeit some of their burners had shades instead of being left unshaded in the more conventionally "Gothic" manner. These brackets were "decorated" in gilt, red, and blue. The rather prickly silhouettes of the ivy leaves suited the taste of the 1870s for stylized outlines. This plate is one of a series of 110 now in The Metropolitan Museum of Art, all but the last three of which are lithographed in color. Among the lithographers are Brett Fairchild and Company, Brett and Company, and Schumacher and Ettlinger. The plate numbers run at least as high as 262, so there may well have once been additional chromolithographs now lost. The extant plates include illustrations of such less frequently encountered types of fixtures as "toilets" (chandeliers suspended from brackets for use at dressing tables), cigar lighters, and reflectors as well as the more often illustrated chandeliers, brackets, lamps, pendants, and pillars.

The previous history of the firm that did business under the name of the Archer and Pancoast Manufacturing Company from 1870 until it ceased operations in 1900 has already been traced (see plates 21 and 45 of this report). During the 1860s and 1870s it was certainly the principal New York rival of Mitchell, Vance and Company, and in 1876 seven of its fixtures were illustrated in an article that described the firm as follows: "Archer and Pancoast M'F'G Co., Designers, and Manufacturers of Gasaliers, Candelabra, Artistic Bronzes, Etc . . . one of the largest and most popular manufacturers of this class of goods . . . has grown to immense proportions."[133]

Courtesy of The Metropolitan Museum of Art, The Elisha Whittlesey Fund, 1951.

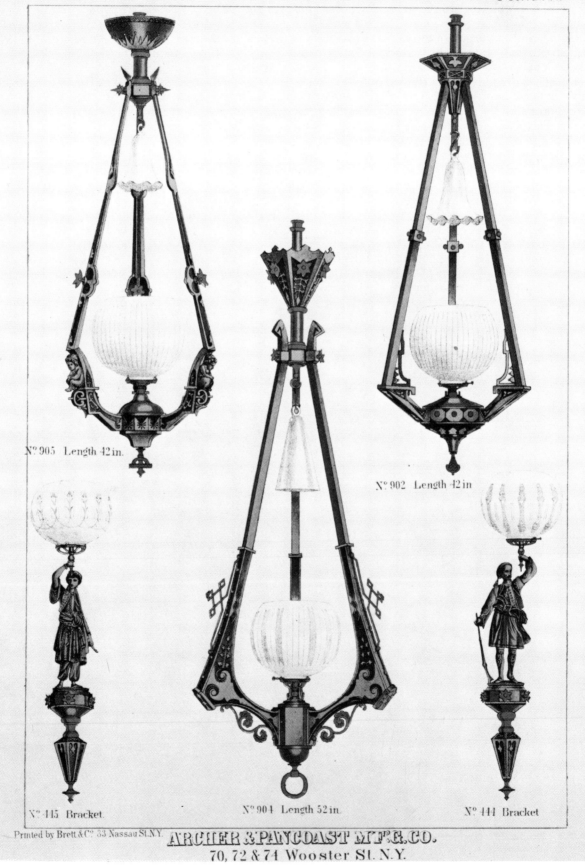

Nº 905 Length 42 in.

Nº 902 Length 42 in.

Nº 415 Bracket.

Nº 904 Length 52 in.

Nº 444 Bracket

Printed by Brett & Cº 33 Nassau St. N.Y.

ARCHER & PANCOAST MF'G. CO.
70, 72 & 74 Wooster St. N.Y.

Plate 82

Pendants and brackets from Archer
and Pancoast Company catalogue, probably 1876.

These angular bronze-finished Archer and Pancoast hall pendants clearly show the reaction against the earlier Neo-Rococo style that came about during the 1860s and continued during the 1870s under the stylistic misnomer, "Eastlake." Nothing in these designs can be called Neo-Grec. The pendant at the upper left, no. 905, is decidedly eclectic. It has the angularity of the Eastlake manner, combined with putti (in a rather disconsolate looking crouch) who look like refugees from the earlier, more romantic style of the 1850s.

The pendants vary in length from 42 to 52 inches. Note the smoke bells and the single set-screws securing the shades. A hall pendant similar in general style to these hangs in the hall of the restored Mark Twain House in Hartford, Connecticut. That house was originally completed in 1874.[134]

The two brackets seem also to refer back to an earlier romanticism. One of the figurines appears to be a Greek evzone, perhaps Marco Bozzaris, the hero of Fitz-Greene Halleck's once famous poem of that name. The other is in female oriental garb, possibly representing Thomas Moore's Lalla Rookh.

Plate 224.

No. 1502.
Length 60 in.

7 Lts.
Spread 34 in.

No. 890. 1 Lt.
Extends 11 in.

No. 1490.
Length 52 in.

3, 4, 5 & 7 Lts.
Spread 26 in.

MANUFACTORY AND WAREROOMS

Archer & Pancoast Mfg. Co.

Nos. 70,72 & 74 Wooster Street, NEW YORK.

Center slide chandeliers and bracket
from Archer and Pancoast Company catalogue, probably 1876.

Plate 83

The Archer and Pancoast plate no. 224 shows two chandeliers and a single bracket finished in bronze with touches of gilding. The Neo-Grec center-slide chandelier at the left could have been intended for either a library, a dining room, or a back parlor. It has none of the specific symbolism that signified "appropriateness" as that term was understood by style-conscious Victorians. The chandelier at the right has minor details that relate to the Eastlake manner as it was interpreted commercially in America. It was almost certainly intended for use in a dining room, since the stag's heads above the branches were meant to represent edible game. The left-hand chandelier measured 5 feet high by just under 3 feet wide, and the one at the right was 4 feet 4 inches high with a spread of 2 feet 2 inches. Presumably the heights were measured with the center slides closed. Note that all burners were counted when describing center-slide fixtures: thus, a "seven-light" center-slide fixture had six branches plus the central burner.

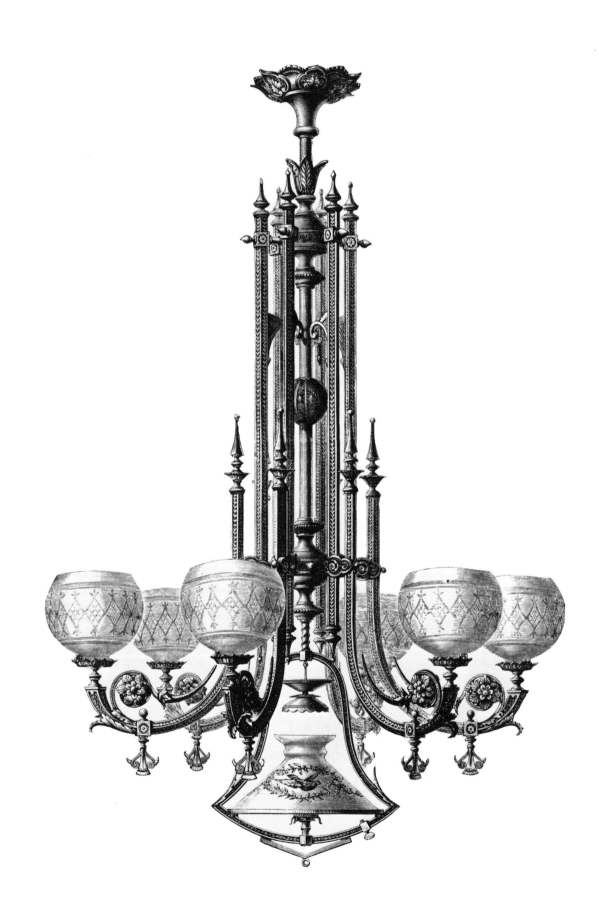

Plate 84

Engraving of center-slide chandelier
by Archer and Pancoast Company, 1876.

Competition in the manufacture of center-slide chandeliers was evidently keen. Mitchell, Vance and Company boasted of the "superiority over all others" of their "patent double slide centre light chandelier," while Archer and Pancoast dubbed the "extension centre-light attachment" they patented in May 1874, with the inspiring name "Excelsior." This illustration of a chandelier incorporating an "Excelsior" center-light was published in London by the *Art Journal* in 1876.[135] The accompanying text under the heading "American Art-Manufactures" reads as follows:

> We present a design for a gas-chandelier selected from the exhibition-rooms of Messrs. Archer, Pancoast and Company, of New York. It is in the style of the time of Louis XIV [a stylistic attribution that would have astonished the Sun King], and is intended for the drawing-room or library. With the extension centre-light attachment, which is known as the "Excelsior," and patented under that name in May 1874, it is also especially adapted for use in the dining-room. The attachment admits of the lowering of the centre-light, and argand burner, from the main body of the chandelier to any desirable distance. The mechanism of the attachment is plain and simple in construction, and its operation is free from many of the intricate contrivances peculiar to slide-chandeliers as heretofore made.... The general effect of the chandelier is light and graceful and yet the central standard renders it unusually strong and massive. The ornamental work is of the finest workmanship, and the whole is richly gilt. It was designed by Mr. J. F. Travis.[136] The centre-light attachment was awarded a silver medal at the recent fair of the American Institute; and a similar medal was also given to the firm for the superior quality of their work....

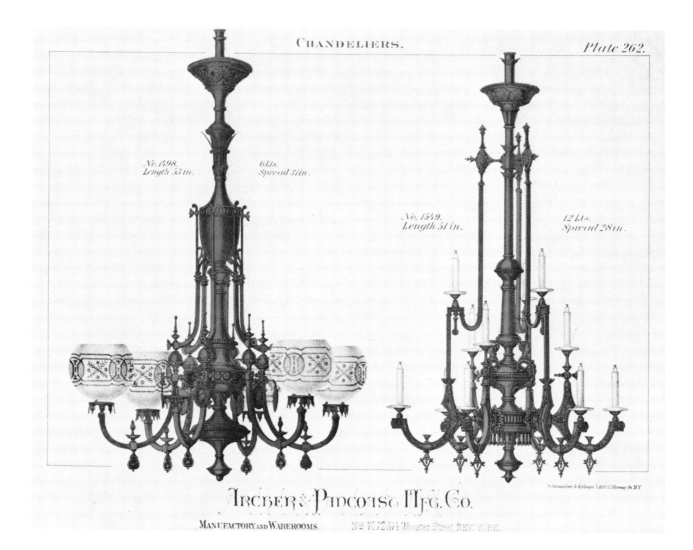

No. 1498.
Length 53 in.

6 Lts.
Spread 31 in.

No. 1549.
Length 51 in.

12 Lts.
Spread 28 in.

Archer & Pancoast Mfg. Co.

MANUFACTORY AND WAREROOMS. Nos. 70, 72 & 74 Wooster Street, NEW YORK.

Plate **85**

Chandeliers from Archer and
Pancoast Company catalogue, probably 1876.

Plate no. 262 in the Archer and Pancoast series shows two gilded chandeliers probably designed for drawing or reception rooms. Possibly the example on the left would have been called "in the style of the time of Louis XIV," and it is probable that the one on the right would have been so described in the 1870s. The age of accurate copying of past styles had not yet arrived; free interpretation of historic modes was still the prevailing practice. The six-light chandelier was 53 inches high and had a spread of 31 inches. The two claws opposite the set-screw can clearly be seen engaging the bases of the two right-hand shades, a detail that is rarely illustrated. The 12-light gas candle chandelier at the right was 51 inches high and 28 inches wide.

The right hand chandelier is one of the very few Archer and Pancoast fixtures shown with gas candles. None of the 92 fixtures of all varieties shown in the Cornelius and Sons catalogue has gas candles, and only one (a church chandelier) of the 12 fixtures illustrated in the Mitchell, Vance and Company *Centennial Catalogue* (really more a brochure than a full-fledged catalogue) has them. When gas candles were used during the 1860s and 1870s, they were most often in dining rooms. Had this chandelier been intended for a dining room, however, it would probably have had a center-slide light. Aside from their use in church fixtures, gas candles appear to have been reserved almost exclusively for dwellings of some consequence until nearly the end of the century. Then they were often used on combination gas and electric fixtures.[137] In this plate the shape of the burners and the glass bobeches and porcelain or opaque glass "candle" sleeves can be clearly seen.

Courtesy of The Metropolitan Museum of Art, The Elisha Whittlesey Fund, 1951.

All the latest and most improved styles of Chandeliers and Reflectors are constantly on hand and made to order.

In ordering Reflectors or Chandeliers, send size of room to be lighted, and state if they are to be inserted **IN** or suspended **FROM** ceiling.

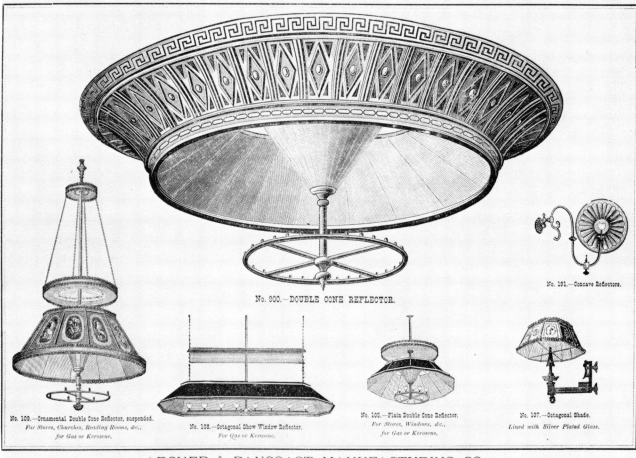

No. 900.—DOUBLE CONE REFLECTOR.

No. 191.—Concave Reflectors.

No. 109.—Ornamental Double Cone Reflector, suspended.
For Stores, Churches, Reading Rooms, &c.,
for Gas or Kerosene.

No. 108.—Octagonal Show Window Reflector.
For Gas or Kerosene.

No. 105.—Plain Double Cone Reflector.
For Stores, Windows, &c.,
for Gas or Kerosene.

No. 107.—Octagonal Shade.
Lined with Silver Plated Glass.

ARCHER & PANCOAST MANUFACTURING CO.,
70, 72 & 74 Wooster Street, New York.

Plate **86**

Reflectors from Archer and Pancoast Company catalogue, probably 1876.

The last three uncolored plates in the Archer and Pancoast series illustrate reflectors. Patents for reflectors were registered as early as 1860 (by another manufacturer, Isaac P. Frink), but the devices seem to have reached their greatest proliferation during the 1870s. They were essentially a sideline of Archer and Pancoast's business, but the firm obviously felt that aspect of the trade sufficiently worthwhile to warrant securing patents based on minor variations of the basic patents secured by another manufacturer. Reflectors were lined with either mirrored glass or silvered metal and were used wherever intense, concentrated light was required. They were made in various sizes, depending on the area to be illuminated. Note that Archer and Pancoast's customers were instructed as follows: "In ordering Reflectors or Chandeliers, send size of room to be lighted, and state if they are to be inserted In or suspended From ceiling." The large picture, "No. 900 – Double Cone Reflector" appears to have been designed for insertion in a ceiling. Presumably inserted fixtures were connected with vents to draw off the heat. The "Ornamental Double Cone Reflector" at the lower left was intended "For Stores, Churches, Reading Rooms, &c." and could be fitted for either "Gas or Kerosene," as could the "Octagonal Show Window Reflector" and the "Plain Double Cone Reflector" adjoining it. The octagonal shade on the bracket at the lower right was "Lined with Silver Plated Glass." The show window reflector was used in connection with an eight-burner gas "T" and met a comparatively new demand engendered by the rapidly increasing use of large sized plate glass in store windows.[138]

Yale classroom with reflector chandeliers.

Plate 87

The classroom used by Professor Othniel Charles Marsh (1831-1899) at Yale was lighted by what appear to be Archer and Pancoast's Plain Double Cone Reflectors (cf plate 86). The pioneer paleontologist's students evidently required more light than the stained glass windows of the Peabody Museum lecture room could admit in order to take their notes. Observe that the striations of the reflecting surfaces of the shade are at right angles to those of the inner cone, a device to increase the reflective power of the light.

A reconstruction of reflectors used
at the Centennial Exhibition, Philadelphia, 1876.

Plate 88

The manufacturing business established in 1857 by Isaac P. Frink specialized in making reflectors. Between April 10, 1860, and April 3, 1883, Frink registered at least 13 patents for reflectors.[139] The I. P. Frink brochure for 1883, referred to "Frink's Patent Reflectors for gas, kerosene, electric, or day-light" and called his device "the Great Church Light for churches, halls, theaters, depots, stores, and public buildings generally." Frink's improved silver-plated corrugated crystal glass reflectors were praised as the best reflectors available. Because gas was listed first as a source of light, it seems safe to assume that most, or at least a majority, of the Frink reflectors were fitted with gas burners. The brochure concludes with a list of 406 presumably satisfied customers and an analysis of that list provides an excellent indication of the uses of most of the reflectors. The list included 243 testimonials from churches, most of them nonliturgical in their form of worship; for instance, 111 were from Methodist churches. Churches using older forms of liturgy tended to favor the Gothic Revival styles.

Theaters formed the next largest building category and together with halls numbered 51. Among the 15 government buildings listed were the statehouses of Georgia, Illinois, Mississippi, Ohio, and Pennsylvania, as well as the Canadian Houses of Parliament in Ottawa. Among the structures furnished with I. P. Frink reflectors by 1883 were: railroad stations (including Grand Central Depot in New York City), steamship piers (including the French Line and Inman Line), market buildings, armories, locomotive works, and other factories, exhibition halls (including Mechanics' Hall in Boston), business and commerce buildings (including the New York Life Insurance Company and Tiffany's store), and the picture galleries of the Boston Museum of Fine Arts, The Metropolitan Museum of Art, and the earlier Century Club building in New York.

The fixture shown here is an electrically lighted reconstruction by the Smithsonian Institution of the type of fixture used to light the main exhibition building at the 1876 Centennial Exposition in Philadelphia. The originals, of course, were gaslighted and the reconstruction, made for the Smithsonian exhibition titled "1876 A Centennial Exhibition," is based on photographs. The original fixtures appear to have been Frink reflectors, although the Centennial is not listed among the installations in the 1883 Frink brochure.

One use of reflectors has yet to be mentioned — the illumination of billiard tables. A fixture with three reflecting shades may be seen over the billiard table at the restored Mark Twain House in Hartford, Connecticut.[140]

Milwaukee Department Store Interior, ca. 1870.

Plate **89**

This detail from a W. H. Sherman stereograph of Chapman's Dry Goods Emporium in Milwaukee, Wisconsin, shows a typical four-light chandelier ca. 1870. Another view of the same store shows one of the long aisles leading to the dome lighted by at least six chandeliers matching this one. Although reflectors might have provided more efficient light for shop interiors, it is probable that they did not conform to the public's notion of elegance. Views of store interiors made during the 1870s almost invariably show more or less ornamental chandeliers. It therefore seems likely that the use of reflectors in commercial emporia was confined to the display windows.

From the author's collection.

FRANK LESLIE'S
ILLUSTRATED
NEWSPAPER

Entered according to the Act of Congress, in the year 1877, by FRANK LESLIE, in the Office of the Librarian of Congress at Washington.

No. 1,157—Vol. XLV.]   NEW YORK, DECEMBER 1, 1877.   [PRICE, 10 CENTS.

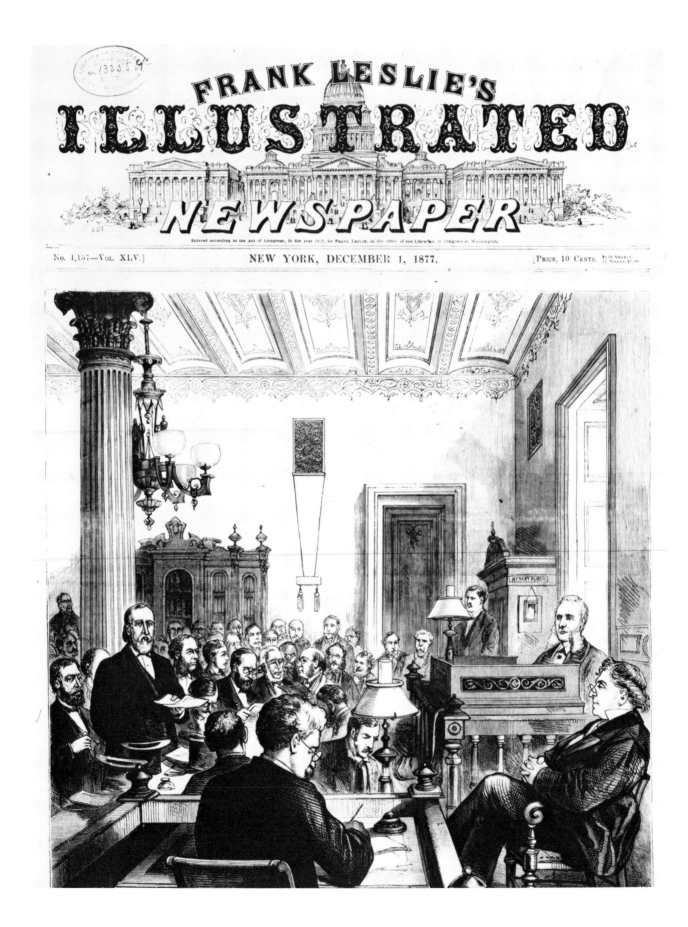

The chandeliers and brackets used in public buildings did not differ in design from those made for residential use, except for an occasional bit of symbolism, unless great size was called for, as in the case of theater auditorium chandeliers.

This wood engraving from *Frank Leslie's Illustrated Newspaper* for December 1, 1877, shows the interior of the Surrogate's Court Room in New York City as it appeared on November 14, 1877, during the famous case contesting Commodore Cornelius Vanderbilt's will.

Note that the chandelier looks much as expected of any six-branched ornate fixture of the period, with the possible distinction that the figurines above the bowl may have represented such abstractions as justice, prudence, law, etc. There is evidence in this picture of specialized use, i.e., the gaslamps on the judge's bench and the clerk's desk. The shades may or may not be reflectors, but the glass chimneys certainly indicate Argand burners, which gave considerably more light than the standard fishtail or batswing burners. Frequently desks, in public rooms where much writing was required, were furnished with gas Argand lamps.

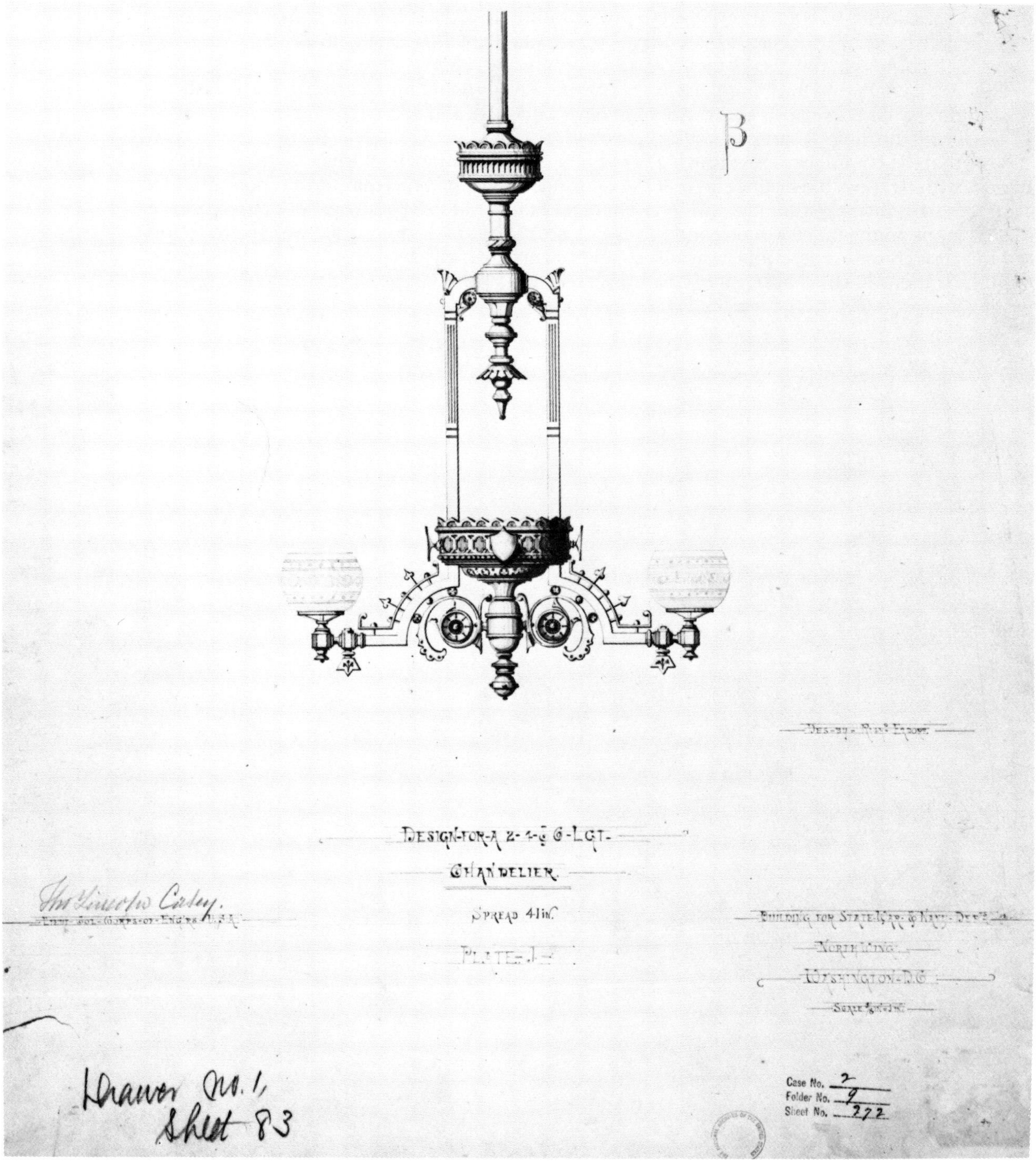

B

DESIGN FOR A 2 & 6 LGT.

CHANDELIER.

SPREAD 41 in.

PLATE. I

BUILDING FOR STATE WAR & NAVY DEPTS.

NORTH WING.

WASHINGTON D.C.

Jno Lincoln Casey.

Drawer No. 1
Sheet 83

Plate **91**

Von Ezdorf design for chandelier, State, War and Navy Building, 1876-1886.

The custom of designing gas fixtures specifically for important U.S. Government buildings continued until the end of the gas era. At least three drawings for lighting fixtures by Richard Von Ezdorf (1848-1926), the artist who designed most of the ornamental work in the former State, War, and Navy Building (now the Old Executive Office Building), have survived.[141] The Ezdorf design seen here is for a chandelier having two, four, or six lights. Its spread was 41 inches. The shields on the fixture, blank in this drawing, were to bear the insignia of the particular military personnel occupying the office, e.g., Army Engineers, Artillery, etc. A supplementary drawing lettered "B" shows four alternative shield designs. This chandelier has the decidedly stiff and rather spiky "Eastlake" character so congenial to the taste of the 1870s. It was designed between 1876, when Ezdorf was transferred to the War Department rolls, and 1886, when the need for new designs was filled.[142]

Another Ezdorf design is for a chandelier with two alternative designs for branches. Sea horses and sea monsters were included among the ornaments. Perhaps the most striking Ezdorf lighting fixture drawing is the one for brackets that are still extant in the former Navy Library, the so-called "Indian Treaty Room." They have been likened to nautical figureheads, an appropriate simile for a Navy Department room. The brackets are in the form of winged half figures of children symbolizing peace, war, liberty, science, industry, and other abstractions above which are horizontal bars holding three burners each.[143]

On September 14, 1882, a package of 41 photographs addressed to George H. Elliot, Lieutenant Colonel of Engineers, was received in the Office of the Chief of Engineers, U.S. Army, in Washington. The photographs, some of them "phototypes" by F. Gutekunst of Philadelphia, were from Thackera, Sons and Company, "Manufacturers of Gas Fixtures, Bronzes, &c." of 718 Chestnut Street in Philadelphia and illustrated 60 of their fixtures.[144]

Many of the fixtures illustrated in the 1882 package were obviously quite new, as the mounts to which they are affixed are stamped (not printed) "Thackera, Sons & Co." Others can be dated before 1882, as their mounts are printed "Thackera, Buck & Co." All, however, appear to have been recently designed, as they show a marked departure from the Neo-Grec and misinterpreted Eastlake mannerisms current in 1876 and thereafter. It should also be noted that at this time all the shades, without exception, exhibited at the Centennial were of the small-necked variety, whereas each shade in the Thackera photographs is of the new wide-based type. That significant change therefore occurred between 1876 and 1881 at the latest.[145] The improved shade design, which greatly reduced or entirely eliminated flickering, was rapidly adopted, and many older fixtures were fitted with new shade holders to accommodate wide-based shades.

The earlier association of Benjamin Thackera with William F. Miskey and William O. B. Merrill in 1866 was discussed in plate 45 of this report. In 1871 or 1872, Thackera, Buck and Company was founded with Benjamin Thackera, William J. Buck and John H. Southworth as partners. By 1877, Charles Thackera and Byron H. Buck had joined the partnership, which continued to operate as Thackera, Buck and Company until sometime in 1881. In 1882 the firm was listed as Thackera, Sons and Company and did business under that name until sometime in 1887. The partners were Benjamin Thackera, John H. Southworth, Charles Thackera, and Alexander M. Thackera, until 1887 when Southworth left the firm. From 1888 until the turn of the century or later the firm was styled the Thackera Manufacturing Company. In 1900 the listing read "Thackera Manufacturing Co. Gas Fixtures 1606 Chestnut Street."[146]

The fixtures in the photographs are for the most part very *avant garde* for their date. They show a sophisticated awareness of artistic developments in England, where William Morris's reformist principles, Walter Crane's elegant stylizations and Edward W. Godwin's Anglo-Japanese designs (as well as the work of many others) were revolutionizing aesthetic perceptions. Many of the Thackera chandeliers have a well-proportioned lightness and elegance of design that was exceptional in American commercial work around 1880.

The three hall pendants shown here are typical Thackera designs of their period. Naturalism has been wholly abandoned in favor of stylization. The flowers and bird etched on the cylindrical shades are rendered in the Anglo-Japanese manner and were done at least three years before 1885, when Gilbert and Sullivan's *Mikado* created a popular rage for *japonisme*. The sunflowers and other metalwork motifs of the left-hand pendant are derived from the English aesthetic movement, and the utter simplicity of the restrained fixture in the middle may owe something to William Morris's reforms. It depends entirely upon its subtle proportions for its handsome effect.

Of the 15 pendants illustrated in the photographs, three are of the square lantern type with clear glass sides, and three others, two of them square, are hung with spear prisms. One pillar and three standards may also have been intended for hall lights, as they are suitable for mounting on newel posts.

Chandelier by Thackera, Sons and Company, 1882.

Plate 93

The photographs, or "phototypes," sent by Thackera, Sons and Company to Lieutenant Colonel Elliot in 1882 were not accompanied by any list of customers, satisfied or otherwise, as was common advertising practice, nor has any such list so far come to light. It is therefore difficult to ascertain the status of the firm in the trade as a whole, but the quality of their designs suggests that their standing was high among their more discriminating patrons. It is known that in 1876 Thackera, Buck and Company supplied the gas fixtures for Horticultural Hall, one of the two major permanent buildings erected for the Centennial Exposition.[147]

The restrained and elegant fixture on this plate is an example of Thackera work using tubular forms almost exclusively. The rectangularity so characteristic of design during the 1880s is not apparent here, but is found in various other Thackera designs as will presently be seen. This four-branched chandelier measured 49 inches high by 27 inches wide. Two of the burners and their accompanying shades and shade holders have been removed to show the design more clearly. Note that the shades are cut as well as partially frosted, thus providing brilliant refraction. Cut glass shades were more commonly used on crystal chandeliers of the period than on brass ones. They were among the rare and most expensive shades made, along with hand-painted ones.

Courtesy of the Office of the Chief of Engineers, Photo No. 77-F-156-44-863 in the National Archives.

195

Chandelier by Thackera, Buck and Company, ca. 1881.

Plate 94

In the Thackera photographs, 19 of the 60 fixtures, or just one less than a third, are ornamented with ceramic ornaments in the Japanese taste. Three of the four standards shown have fairly large vases, but most of the ornaments, like the one seen here, are relatively small. This small chandelier combines Eastlake metalwork of very flat profiles, angular outlines, and square sections with its Anglo-Japanese ceramic baluster. This eclectic combination of neomedieval and oriental forms is, however, blended to form a harmonious whole. The one element inconsistent with the design is the Neo-Grec griffin and rinceau pattern etched on the shades. That pattern was evidently very popular, as it appears on three other sets of shades in the Thackera photographs, as well as on the shades of two fixtures previously discussed (see plates 24 and 63 of this report). This three-branched fixture (one of the shades has been removed with its holder to clarify the illustration) was 3 feet high and 2 feet wide. "Thackera, Buck and Co.," is printed on the mount to which the photograph is affixed, indicating that this fixture must be dated no later than 1881.

Courtesy of the Office of the Chief of Engineers, Photo No. 77-F-156-44-890 in the National Archives.

197

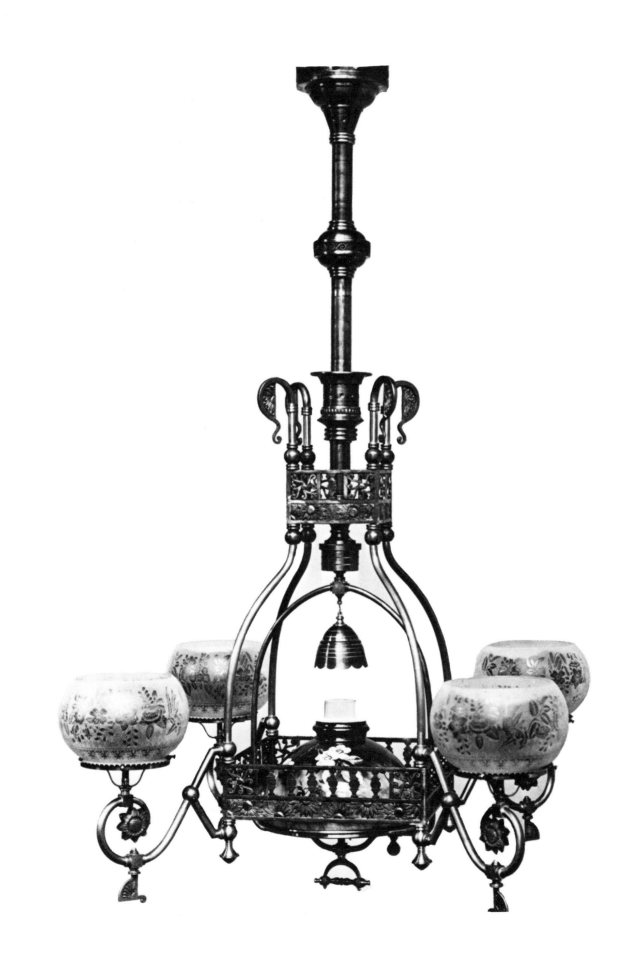

Chandelier with center slide, Thackera, Sons and Company, 1882.   Plate **95**

If the 33 chandeliers in the Thackera photographs, 8 (like this one) have center slides. Representing just under 25 percent of the total, this may indicate a waning popularity for center slides, which apparently tended to get out of order, judging from the many patent gadgets that claimed superiority over other presumably more troublesome models. This fixture was probably designed for a dining room or library. The stylized floral motifs, fan-like patterns, and spindle work gallery of the metalwork combine Anglo-Japanese and Eastlake forms. The square gallery surrounding the slide fixture and its painted shade is reminiscent of similar spindle work found on furniture of the period. The size of this chandelier was 52 inches high with a 34 inch spread.

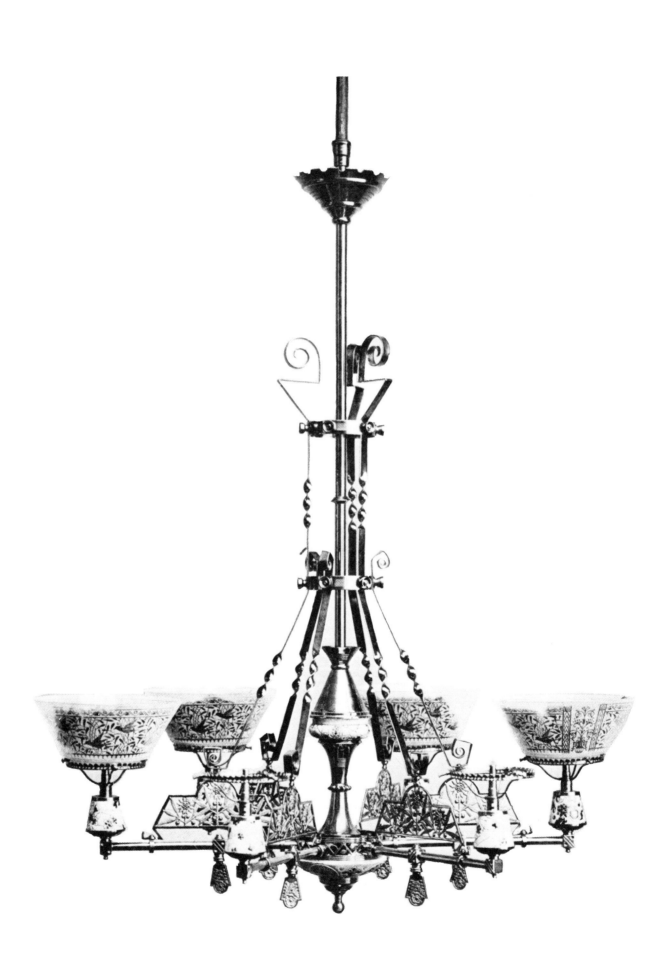

Few if any fixtures designed around 1880 in America were as up-to-date in their conformity with advanced artistic principles as this one from the Thackera collection. Two of the shades have been removed to show the design more clearly. The thin, twisted and spiraled brass trim and the cast-brass floral motifs, combined with the ceramic sections under the burners and decorating the stem, are more or less consistently in the Anglo-Japanese manner. Etched with a birds-in-bamboo pattern, the shades are most appropriate to the total design. This six-branched fixture measures 41 inches high by 27 inches wide.

The Thackera photographs may give a clue to the prevalence of various types of fixtures around 1880. Thirty-three of the 60 fixtures shown on the 41 pictures were chandeliers. Four were standards. Two were toilets, that is, small chandeliers on cantilevered brackets; and the rest were hall or two-branched pendants. As already noted, slightly less than a quarter of the chandeliers had center slides. Almost a third of the chandeliers and other fixtures (as noted) had ceramic ornaments in the Japanese manner. The most vital new design influence of the moment was therefore accorded ample recognition.

Eight-branch, glass ornamented chandelier, Thackera, Sons and Company, 1882.

Plate 97

Only six Thackera chandeliers out of the 33 in the photographs have any glass ornaments at all, and those six that can in any way be classified as "crystal" do not resemble the more conventional glass examples previously discussed. Several have a far greater proportion of metal to glass than was formerly customary. Some have stems ornamented by thin brass arches studded with glass spear points. Others combine two types of prism in a single fixture. All are, to some degree, new and original. A Thackera, Buck and Company two-tiered fixture with notched spear prisms and a few glass bells in the form of fuchsia blossoms appears to have had silvered metalwork and was somewhat more conventional than the others.

The eight-light fixture illustrated here (four shades have been removed) is the most original of the glass-ornamented Thackera fixtures. Each annular element is fringed with glass balls, a design motif perhaps suggested by the textile ball fringe so popular for use on furniture and lambrequins during the 1870s and 1880s. So little glass is used, however, that it is moot whether or not this creatively inventive design should be termed a "crystal chandelier." This fixture was 4 feet high and had a spread of 26 inches.

It is worth noting that as early as 1882 at least one crystal chandelier was already imitating a late 18th-century English type. Thackera's no. 9560, a six-light example, had a base of concentric rings of prisms and cascades of faceted glass bead chains that descended from a crowning ring and flared slightly outward to the ring supporting the burners. The center stem was thus entirely concealed by the glass "basket." That style of chandelier, popular during the Federal period and even later, was very probably considered by ca. 1880 to be "Colonial."

Courtesy of the Office of the Chief of Engineers, Photo No. 77-F-156-44-953 in the National Archives.

203

Plate **98**

Six-light glass ornamented chandelier, ca. 1880.

This six-light chandelier, 6¾ inches taller than the eight-light one on the previous plate, is now in The Metropolitan Museum of Art. It originally hung in the dining room of William Ryan, a prosperous Dubuque, Iowa, meat packer who had the inelegant nickname of "Hog Ryan." One year during the 1880s Mr. Ryan's daughters had the family's dining room redecorated with new wainscoting, William Morris wallpaper, and this chandelier in the latest aesthetic taste. The flat, chrysanthemum pattern of the metalwork panels in the upper zone resembles similar panels shown on a chandelier, no. 720, in the I. P. Frink catalogues for 1882-1883, and on that basis this chandelier has been attributed to Frink.[148] However, the even closer resemblance between it and the Thackera chandelier on plate 97 should also be noted.

Courtesy of The Metropolitan Museum of Art, Edgar J. Kaufman Charitable Foundation Fund, 1969.

INCANDESCENT GAS LIGHT
(NEW & IMPROVED SYSTEM)
ELECTRIC LIGHT SURPASSED.

All the advantages of Electric Light and none of its drawbacks.

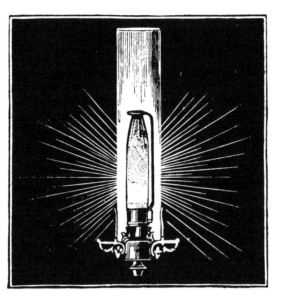

Facsimile of the first advertisement of the Welsbach Light, which appeared in the "Journal of Gas Lighting" for December, 1890.

On October 21, 1879, Thomas Alva Edison (1847-1931) succeeded in making an incandescent electric light bulb that burned for over 40 hours. While it is an oversimplification to say that Edison invented the electric light, his improvements on earlier discoveries and his development of a central distributing system for the requisite current made electric light commercially feasible. By 1882, his famous Pearl Street Station's dynamos were supplying direct current over wires to users in New York City. From then on, gaslighting had its first really serious competitor.[149]

Had a successful incandescent gaslighting device not been invented soon thereafter, it is quite possible that gaslight would have been almost entirely supplanted by electric light before the close of the century. However, gas was able to continue an increasingly futile rear guard action until about 1910 in domestic lighting and even a little longer in street lighting, thanks in large part to the Welsbach burner, or gas mantle. The Austrian chemist Dr. Carl Auer von Welsbach (1856-1929), who was raised to the rank of Baron by Emperor Franz Josef for his discoveries, developed his first incandescent gas mantle during the years 1885-1886. It consisted of a cotton mesh cylinder gathered at the top and treated with a solution of rare earth minerals (principally thorium dioxide with a touch of cerium oxide). After the cotton was burned off, the incombustible oxides formed an open cone that glowed with intense light when placed over a burner within a glass chimney. The early mantles were fragile, but in 1887, when the first factory for their manufacture was established, a life of from 800 to 2,000 hours was claimed for them.[150]

Contemporary accounts reported:

> The electricians will have to wake up, if they do not wish to be left behind in one of their most important fields of labor. Within the last few days a new light has made its appearance in the City [of London] which rivals in brilliance the best they have yet produced.... Dr. Welsbach calls this his "mantle".... The moment the flame is applied the mantle becomes incandescent, that is, rises to white heat, and gives out a brilliant, mellow light, which it may be said, without any exaggeration, will compare favourably with any electric light yet put on the market....[151]
>
> The new incandescent gas light invented by Dr. Carl Auer von Welsbach, of Vienna, promises to confer a real boon upon all who use gas as an illuminant. The advantages claimed for this invention — and not only claimed, but proved by actual experiment — are manifold.... The new system has been proved both practically and commercially. It has been in use for some time in Vienna, Berlin, &c., and it is considered the greatest discovery made in gas lighting since gas was first introduced for illuminating purposes.[152]

The advertisement reproduced in facsimile on this plate was the first to appear for the Welsbach burner and was published in the *Journal of Gas Lighting* in London in December 1890. Note that the text boasted "Electric Light Surpassed." The "drawbacks" of electric light to which the advertisement referred were primarily the need for stringing wires and the fairly frequent power failures. The rapid spread of incandescent gaslight is indicated by the sales statistics for Great Britain: 20,000 Welsbach burners were sold in 1893, 105,000 in 1894, and 300,000 were sold in 1895.[153]

Reprint from Chandler, *Outline of History of Lighting*, Courtesy of American Gas Association, Inc. Library, Arlington, Virginia.

This photograph taken around 1906 by the noted architectural photographer Frances Benjamin Johnston (1864-1952) shows the parlor in the house of the famous reformer and women's suffrage advocate Susan Brownell Anthony (1820-1906). The small and inexpensive three-branched chandelier appears to date from the 1870s and was evidently inadequate in its original form, as its owner made three improvements. The burners first supplied with this cheap little fixture were no doubt standard iron fishtails. The burner at the left appears to be a governor burner, an expensive type, and probably the shade of the 1880s type replaced an earlier shade when the improved burner was installed. Another improvement, the reflector behind the burner in the background, was designed to throw a more adequate light upon the painting above the mantelpiece. The third and most important improvement was the installation of the Welsbach burner in the foreground. Note the chains hanging from the little bar that controlled the flow of gas to the mantle. Often such chains are mistaken for electric light pulls when seen in old photographs. When two are visible as close together as in this photo, it indicates that they were used with a gas mantle. The crimped white glass shade helped to direct the light downward, where it was most needed.

As was the case with the introduction of wide-necked shades, so it was with Welsbach burners, i.e., many older fixtures were updated with the new and more efficient type of burner. By the turn of the century gas mantles were commonplace, although many flat flame burners were still being made. A photograph taken in 1902 of the Boston studio occupied by the prominent designer-craftsman Isaac Elwood Scott (1845-1920) shows Welsbach burners very similar to the one used by Miss Anthony.

A picture of a New York City barber shop taken in 1903 by the photographic firm headed by Joseph Byron shows lighted Welsbach burners and electric filament bulbs in combination on the same fixtures.[154]

1880s chandelier with inverted gas burner
and light controls, post 1905.

Plate 101

A technical report written in 1908 describes an innovation in gaslighting.

> The latest modification of incandescent gas lighting, introduced in 1905,
> consists in the use of inverted burners. The inverted gas burners throw the
> light downward and hence do not cast shadows; the mantles are shorter
> and hence do not break quite so easily; the light given off is better and stronger,
> and the gas consumption is less. For these reasons the inverted incandescent gas
> lamp has quickly proven successful, and its popularity is bound to increase. . . .[155]

That passage had a sanguine note about it, predicting increasing popularity for inverted
incandescent burners. Their popularity did indeed increase among gas users, but by
then electricity was advancing so fast in popularity that it was almost too late for any further improvements in gaslighting.

One other effort was made to compete successfully with electric light. In 1899 the Swiss inventor Conrad Adolfe Weber-Marti of Zurich patented "an improved device for opening and closing
low pressure conduits for gas from a distant point of actuation," in other words, a gaslight switch.
That invention was improved by Neville Edwards in 1906 and again in 1909, but there is no
evidence that gas switches were used to any extent in this country.[156]

This chandelier, now in the Smithsonian Institution Furnishings Collection, appears to date
from the 1880s. Two of its four branches (all now electrified) were later adapted for use with
inverted incandescent burners. The mantles were approximately the size of the small electric
bulbs that have replaced them. Note that small glass globes were used, and that chains controlled
the valves admitting gas to the mantles. The amount of light could thus be precisely managed.

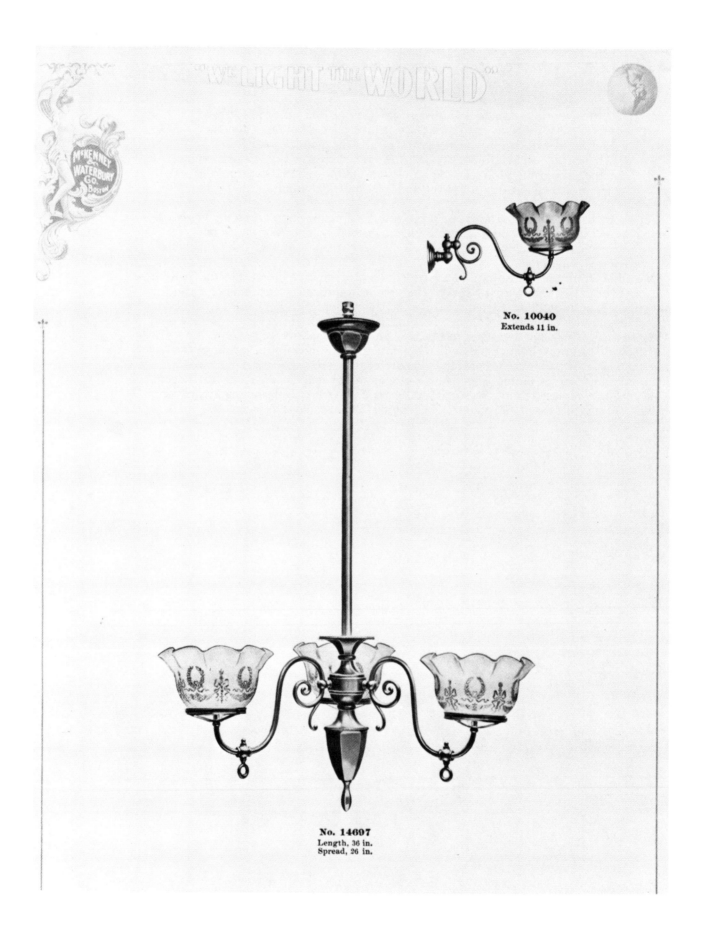

No. 10040
Extends 11 in.

No. 14697
Length, 36 in.
Spread, 26 in.

"Colonial" chandelier and bracket from the McKenney
and Waterbury Company catalogue, ca. 1900-1910.

Plate 102

At an undetermined date, the McKenney and Waterbury Company, a Boston-based firm describing themselves as "manufacturers and jobbers of gas, electric, and combination fixtures," issued their *Gas Catalogue* G. McKenney and Waterbury did not indulge in stylistic nomenclature, although their fixtures freely interpreted past styles, and their boasting was confined to their motto, "We Light the World." The introduction to *Gas Catalogue* G read in part:

> We have compiled in this catalogue a line of new and exclusive designs in the
> medium grade of fixtures, and will be pleased to furnish additional suggestions
> for whatever class of work you desire. We will furnish on inquiry photographs of
> our better goods, covering everything pertaining to gas, electric, or combination
> fixtures.
> Fixtures are furnished in any of the regular finishes, such as Rich Gilt, Old
> Brass, Flemish Brass, and Polished Brass at list prices. Nickel, Verde Green, or any
> Copper Finish, 10 percent extra. Ormolu or any Silver Finish, 20 percent extra.

The three-branched chandelier and the bracket shown here on their plate 27 are typical of McKenney and Waterbury's "medium grade of fixtures." The chandelier was available with two lights at $7.50 or the three at $9.50. The bracket cost $2.75. The finish was "Old Brass and Black," and the dimensions were: height, 3 feet; and spread, 26 inches. The bracket projected 11 inches. The hexagonal shapes with beaded edges, and the wreath-and-beribboned-torch pattern of the shades are characteristic early 20th-century motifs. Possibly the chandelier would have been classed as "Colonial." The pattern of the shades was Neo-Empire.

An unidentified catalogue apparently dating from somewhere between 1900 and 1910 offered gas fixtures in the following "styles": Colonial, Empire, Flemish, Gothic, L'Art Nouveau, Louis XIV, Louis XV, Louis XVI, Medieval, Mission, Moorish, Renaissance, and Rococo. The finishes available were satin gold, old brass, rich gilt, polished brass, and wrought iron. The firm sold exclusively to the trade and boasted that their stock was composed of "original designs, no copies." Any student of the historical styles would concede their point. Very free adaptation, not copying, was their forte. Even their fixtures in the contemporary Mission style would hardly have received any approbation from Gustav Stickley or Elbert Hubbard's Roycrofters.[157]

No. 10080
Extends 10 in.

No. 13631
Length, 36 in.
Spread, 18 in.

Plate 103

Chandelier and bracket from the McKenney and Waterbury Company catalogue, ca. 1900-1910.

The chandelier and bracket on plate 49 of the McKenney and Waterbury catalogue are slightly more elaborate than those on the previous plate of this report. The bracket extended 10 inches and cost $4. The chandelier was 3 feet high and 18 inches wide. The finish was "English Gilt and Matt." With two lights it cost $11; with three it cost $13.50. In addition to the hexagonal shapes and beaded edges so typical of the period, these fixtures have very late neo-classical versions of the palmette motif. Possibly they would have been called "Empire" in their day.

No. 116
Length, 42 in.
Spread, 24 in.

Plate **104**

Gas candle chandelier from the McKenney and Waterbury Company catalogue, ca. 1900-1910.

Gas candles were popular around 1900 and afterwards on both gas chandeliers and combination gas and electric fixtures. They were made in various shapes: round, square, hexagonal or spirally twisted as illustrated here.

McKenney and Waterbury's plate 43 shows another example of their "medium grade" fixtures. This chandelier, with five lights was 42 inches high with a spread of 24 inches, and cost $31. With only four lights the fixture cost $26.50. In either size the finish was "Old Brass and Black."

The design is derived ultimately from 17th-century Flemish chandeliers, but at the time of manufacture, this example may have been referred to as either "Flemish" or "Colonial." The fleur-de-lis is a contrived adornment.

No. 13857
Length, 42 in.
Spread, 22 in.

Plate 105

Rococo chandelier from the McKenney and Waterbury Company catalogue, ca. 1900-1910.

The quite elaborate five-light chandelier on this McKenney and Waterbury's plate cost $48 and was about at the top of their medium grade line. The finish was "English Gilt and Matt," and the measurements were 42 inches high by 22 inches wide. The gas candles are of the standard plain round type and have plain glass bobeches. The comparatively lavish use of foliate motifs and the C-curves of the gas keys and mantling of the ball at the base of the fixture suggest that the style might have been called either Rococo or Louis XV. The Rococo elements of this design belong to the late Rococo Revival and differ markedly from the Rococo manner popular during the 1840s and 1850s.

There are in all 146 plates in the McKenny and Waterbury catalogue. Among the types of fixtures shown are brackets, pendants and chandeliers, hall pendants, pillars, and lamps, as might be expected. About half of the fixtures had gas candles. Plates 82-96 show chandeliers with art glass domes, then much used for dining rooms. One plate shows mantles, and another shows acetylene burners. Three plates are devoted to shades, and one shows "wire globes" of the type used where breakage or fire were hazards, as in prison corridors or backstage in theaters. Another plate has "torch and key" lighters[158] and tapers as well as burner saws for clogged batswings and fishtails. The lighters sold for $3 per dozen. The last five plates carry pictures of gas cocks in great variety.

When the catalogue was published, McKenny and Waterbury's address was 181 Franklin Street at the corner of Congress Street in Boston. The firm was founded in 1889 by William A. McKenney and Frank S. Waterbury and was still in business as late as 1919. William A. McKenney was in Boston directories prior to 1889 as a clerk with Charles H. Waterbury and Company, who were also listed as manufacturers of lighting fixtures.

Courtesy of the Avery Library, Columbia University.

GAS FIXTURES

LET US SAVE YOU MONEY WHEN YOU BUY YOUR GAS FIXTURES. All we ask is that you make a selection from this catalogue, and after you have received them, if you are not satisfied that you have saved 50 per cent on your purchase, return them to us at our expense and we will gladly refund your money together with all transportation charges both ways.

THE PROFESSIONAL GAS FITTER, PLUMBER OR MECHANIC will do well to buy his fixtures from us. We can save him money and at the same time furnish him with fixtures that will give satisfaction to his customers. Our exclusive designs

cannot fail to please even the most critical and the fixtures have a rich, expensive appearance.

OUR GUARANTEE. We guarantee that the fixtures are made of the best quality of highly polished brass. We guarantee that these fixtures are well made, strong and durable. Every fixture is tested before leaving our house to insure against leakage. We claim our fixtures are superior in quality, finish and durability and we guarantee them to give perfect satisfaction.

NATURAL GAS AND ACETYLENE GAS TIPS. The prices quoted below are for fixtures fitted complete with tips for artificial or manufactured gas. We can furnish gas fixtures for natural gas at an extra cost of 8 cents for each tip or with acetylene tips for 25 cents each tip.

Swing Gilt Brass Gas Bracket.

Made of excellent quality of brass in ribbed design. Furnished complete with wall plate, pillar, tip, and with one or two arms, either

No. 3K2211

with burner cup or with white crystal glass shade in richly cut design and brass shade holder.

No. 3K2202 Single Swing Gilt Brass Gas Bracket with burner cup.
Price, per ½ dozen, **$2.18**; each **37c**

No. 3K2203 Single Swing Gilt Brass Gas Bracket with white crystal glass shade and brass shade holder.
Price, per ½ dozen, **$3.08**; each **53c**

No. 3K2210 Double Swing Gilt Brass Gas Bracket with burner cup.
Price, per ½ dozen, **$3.48**; each **61c**

No. 3K2211 Double Swing Gilt Brass Gas Bracket with white crystal glass shade and brass shade holder.
Price, per ½ dozen, **$4.45**; each **78c**

One and Two Light Polished Brass Pendant.

Suitable for kitchen, cellar, back hall and other places where a fancy gas fixture is not necessary. All brass tubing nicely polished, neat and well constructed. Furnished complete with burner cup, pillar and gas tip. No globe or globe holders furnished at prices quoted below.

No. 3K2235 One-Light Pendant, 30 ins. long.
Price **49c**

No. 3K2236 One-Light Pendant.
Price.......**54c**

No. 3K2225 Two-Light Pendant, 36 inches long.
Price....**$1.15**

One-Light Fancy Gas Pendant Polished Brass.

One of the neatest and most artistic pendants on the market. It is made of polished brass with fluted cup and fancy carved leaf ornaments and fancy gas stop cock. The pendant is very well constructed, and is an ornament to any room. Furnished complete with brass ceiling plate, pillar and gas tip and in one length only, 36 inches. We can furnish this pendant either with or without a globe and holder.

No. 3K2234 Fancy Gas Pendant, with burner cup. Price.**$1.12**

No. 3K2238 Fancy Gas Pendant, fitted with white crystal shade in very rich looking design, and brass burner plate.
Price.......**$1.23**

Hall Gas Fixture.

A very neat and effective fixture for little money. Made of highly polished brass tubing, in graceful design. Comes complete as shown in illustration with ceiling plate, pillar, gas tip, globe, globe holder and fluted ceiling protector hanging over the light. Length of fixture, 30 inches. We can furnish this light with either opalescent or red globe.

No. 3K2245 Price, with opalescent globe **$1.49**
No. 3K2246 Price, with red globe **$2.08**

OUR LEADER HALL LIGHT
$2 39

NEW DESIGN. VERY ARTISTIC.
A GREAT BARGAIN.

THIS BEAUTIFUL FIXTURE is made of solid brass tubing, very highly polished with heavy weight brass ornamentation. We consider this fixture the greatest value in hall lights ever offered for the money.

A HANDSOMELY EMBOSSED CROWN is fitted around the top of the globe. On either side of the frame, just above the light, a wrought brass spray projects, adding greatly to the graceful design. A fluted ceiling protector hangs over the light, while above it is a center ornament in the form of a brass cup, and above that the brass ceiling plate. The globe is of opalescent glass, fluted in a graceful design cylinder shape. An openwork key to turn on and off gas completes the light.

THIS HALL FIXTURE IS FURNISHED COMPLETE with globe, brass pillar, lava gas tip and ceiling plate, just as shown in the illustration. The fixture, when you receive it, is all complete and ready to screw into ceiling. Bear this in mind. Our price is for the fixtures complete. It includes all the parts.

No. 3K2255 Our Leader Hall Gas Fixture, just as described above. Price........**$2.39**

No. 3K2256 Our Leader Hall Gas Fixture, just as described above, but with transparent red globe. Price...................**$3.18**

"The Star" Gas Chandelier.

Made of highly polished brass with exceedingly attractive ornamentation. Consisting of fancy fluted ball ornament and pear shaped pendant. The curved arms project from this ring and have cast brass ornaments in dull finish. Fitted complete, as shown in the illustration, with beautiful pebbied crystal glass shades, ornamented with cut stars. It is also fitted with brass shade holders, pillar, lava gas tips and brass ceiling plate. We can furnish with two or three lights. Length of fixtures, 36 inches. Spread of arms, 19 inches.

No. 3K2275 Two-light fixture.
Price .. **$2.89**

No. 3K2276 Three-light fixture.
Price.. **$3.88**

Imperial Gas Chandelier.

A well made finely finished chandelier and exceedingly attractive to look at. It has a graceful design, being made of extra fine brass tubing, highly polished, with fluted cup, pear shape ornament. Around the center of the pear shape ornament runs a beaded band, from which project the arms. The arms are in graceful descending curves with solid cast brass pendants in fern design. The delicately etched shades add greatly to the beauty of the chandelier. This fixture can be furnished with two or three lights and comes complete with etched shades, brass shade holders, pillar, lava gas tips and ceiling plate.

No. 3K2280 Two-light fixture.
Price.. **$3.35**

No. 3K2281 Three-light fixture.
Price ..**$4.50**

$3 48 PEERLESS GAS PORTABLE LAMP.

A high grade reading lamp for parlor and library, furnished complete, green shade and tubing. The lamp is all ready to use when you receive it. All that is necessary to do is to remove the gas tip from the fixture and slip in the gooseneck. Then turn on the gas and light it. The price we ask includes all the parts, including burner, which is of best quality, being made of polished brass, high grade cap mantle, opaque chimney globe with air holes, green shade (green outside, white inside), fitted with a heavy green bead fringe to match, and six feet of the best quality of mohair tubing fitted with brass gooseneck. A high grade lamp. We are able to offer our customers a high grade reading lamp, and for those desiring a high grade lamp we believe this outfit will give perfect satisfaction. The stand is made of metal, finished in a rich black color. The metal is highly ornamented and trimmed in polished brass. A lamp of this quality and finish together with all the fittings we give has never been sold before for less than $5.00 and more often at $7.50.

No. 3K2191 Peerless Lamp, complete and ready to light.
Price..............**$3.48**

Gilt Gas Chandelier.

For people desiring a low price fixture this is just what they want. Made of polished brass, corrugated design with brass ball ornament in satin finish and artistic loop design. Furnished with two or three lights. Length of fixture, 36 inches. Spread of arms, 36 inches. Remember the prices quoted below are for the fixture complete with crystal glass globes in rich design, brass globe holders, pillar, gas tips and ceiling plate, all complete and ready to put up.

No. 3K2265 Two-light fixture.
Price.. **$1.78**

No. 3K2266 Three-light fixture.
Price **$2.38**

"Au Fait" Polished Brass Gas Chandelier.

$1 89

The very best low price chandelier made. It is made of polished brass in a simple and classical design, yet at the same time most artistic. It is ornamented with two octagonal shaped balls with beaded centers, just above where the arms meet. The lower ball has a flattened octagonal top with a rounded embossed lower part; and the arm connection and keys are of dull finished brass. Furnished with either two or three lights and is complete with white crystal shades in richly cut design, brass shade holders, pillar, lava gas tips and brass ceiling plate. Length of fixture, 36 inches. Spread of arms, 19 inches.

No. 3K2270 Two-light fixture.
Price, **$1.89**

No. 3K2271 Three-light fixture.
Price, **$2.49**

OUR SPECIAL EXTRA VALUE LEADER CHANDELIERS

EXQUISITE IN DESIGN AND DETAIL.

Solid Heavy Brass Tubing, Highly Polished, Elaborately Ornamented, Satin Finish Trimmings.

$4 18

Do not be deceived by the price we ask for this chandelier. The only thing cheap about it is the price.

WE GUARANTEE THIS FIXTURE to be of the best materials. It is made of heavy brass tubing, very highly polished. Part of the tubing is corrugated, giving the fixture a very rich appearance. The fixture has a fluted ceiling plate and is ornamented with a fluted cup in satin finish surmounting a ball with beaded band and large pineapple shaped pendant. This pendant is extra large and elegant, and has attached to it at the bottom a fluted ball in satin finish brass. From this ball five heavy cast brass leaves reach up and are appliqued on the pendant. A ribbed band of satin finish brass surrounds the pendant, from which band project the arms, which have a graceful downward curve and are ornamented with heavy solid cast brass ornaments in dull finish. The keys are fancy openwork in the same finish.

THE CHANDELIER IS FITTED complete with fluted brass ceiling plate, pillars, best quality of lava gas tips, brass shade holders and beautiful new swell shape silver frosted shades, with embossed edges and angel design, having delicate tracery work etched around the figures. Elegant parlor, library or dining room chandelier.

No. 3K2295 Our Leader Chandelier, with TWO lights.
Price................**$4.18**

No. 3K2297 Our Leader Chandelier, with THREE lights.
Price.................**$5.19**

No. 3K2298 Our Leader Chandelier, with FOUR lights.
Price..................**$6.15**

The text of this page from the Sears, Roebuck and Company catalogue for 1908 is most instructive. The remarkably low prices—ranging from 37¢ each for single-swing, gilt brass gas brackets to $6.15 for the chandelier at the lower right when supplied with four lights—may have reflected the competition of electricity as much as any competition within the gas fixture trade. The best of the brackets, a "double swing gilt brass gas bracket with white crystal glass shade and brass shade holder," cost only 78¢, or $4.85 per half dozen. The unshaded pendants below the swing brackets cost 49¢ with one light, or 54¢ with two. They were "suitable for kitchen, cellar, back hall and other places where a fancy gas fixture is not necessary." The "one-light fancy gas pendant polished brass" below those pendants was described as "one of the neatest and most artistic pendants on the market." With a "white crystal shade in very rich looking design," this 36-inch pendant "made of polished brass with fluted cup and fancy carved leaf ornaments and fancy gas stop cock" cost $1.23. The "very neat and effective" hall fixture at the lower left, complete with "ceiling plate, pillar, gas tip, globe, globe holder, and fluted ceiling protector," cost $1.49 with the swirled opalescent shade, or $2.08 with a red glass globe. The more elaborate hall fixture at top center had a lava steatite tip flat flame burner and cost $2.39 with an opalescent shade, or all of $3.18 with a red one.

The gaslamp had a "cap mantle, opaque chimney globe with air holes, green shade (green outside, white inside), fitted with a heavy green bead fringe to match, and six feet of the best quality of mohair tubing fitted with brass gooseneck." Sears and Roebuck claimed that their $3.48 lamp had never before been sold for less than $5 and had more often cost $7.50. The prices of the five chandeliers that are set in eyecatching heavy display type refer to those fixtures with two lights only. With three lights the price was, of course, higher as a second glance reveals. The cheapest, $1.78 with two lights or $2.38 with three, was made of polished brass with a "brass ball ornament in satin finish" and branches of "artistic loop design." The text describing "Our Special Extra Value Leader Chandeliers" reveals what the seller regarded with pride and, presumably, what the buyer prized. In part it reads:

> It is made of heavy brass tubing, very highly polished. Part of the tubing is corrugated, giving the fixture a very rich appearance. The fixture has a fluted ceiling plate and is ornamented with a fluted cup in satin finish surmounting a ball with beaded band and large pineapple shaped pendant. This pendant is extra large and elegant, and has attached to it at the bottom a fluted ball in satin finish brass. From this ball five heavy cast brass leaves reach up and are appliqued on the pendant. A ribbed band of satin finish brass surrounds the pendant, from which band project the arms, which have a graceful downward curve and are ornamented with heavy solid cast brass ornaments in dull finish. The keys are fancy openwork in the same finish. The chandelier is fitted complete with fluted brass ceiling plate, pillars, best quality of lava gas tips, brass shade holders, and beautiful new swell shape silver frosted shades, with embossed edges and angel design, having delicate tracery work etched around the figures. Elegant parlor, library or dining room chandelier.

Note that these mail order fixtures were supplied to gas fitters as well as to householders. Note also that tips for burning natural gas were available at 8¢ extra per tip and that acetylene tips cost 25¢ extra per tip.

Combination gas and electric fixtures were made from the 1880s until the end of the gas era. This eight-light chandelier in the parlor of a Colonial Revival house near Boston is typical of the combination fixtures made around 1900, when the photograph by C. H. Currier was taken. Curiously, the four gas burners are not all identical. The one in the right foreground has a governor below its flat-flame (probably fishtail) burner, and a single pull chain is visible hanging from the gas burner in the left background. The shade of that burner has been replaced by one that does not match the others, presumably after accidental breakage. Because the four electric lights have no pull chains, there must have been a wall switch, although none shows in the photograph. (The button beside the wide opening into the stairhall rang a call bell to summon a servant). The pierced repousse ornamentation of the stem, or "pillar," and the pendant, as well as the openwork keys, scrolled arms with foliate terminations, and fluted canopy, or "ceiling plate," are all characteristics of fixtures dating from around the turn of the century.

Combination gas and electric "T," ca. 1900.

Plate **108**

This charmingly genteel dancing class, now so evocative of nostalgia (although the boys appear to be anything but happy in their pumps and Eton jackets), was photographed at the Staten Island Women's Club by the highly gifted amateur photographer Alice Austen (1866-1952) around 1900. The simple combination fixture at the right has the familiar "T" form so often used where plain, utilitarian lighting was desired. That type of gas and electric fixture was very widely used during the 1890s and early 1900s in schools, hospitals, and the working areas of public buildings, or wherever efficient but unornamented lighting sources were appropriate. The rather small crimped white glass shades of the electric bulbs are typical of such fixtures, as are the plain white gas globes of the kind called "opal." The opal shades diffused the light evenly and softened it agreeably.

At the other end of the spectrum from the simple combination fixture seen in the previous plate is this very elaborate silver-finished gas and electric chandelier in the dining room of the brewer Christian Heurich's mansion (now the Columbia Historical Society) in Washington, D.C. The Heurich Mansion was built in 1892 and furnished in a lavish, if rather unsophisticated, manner. The chandelier, like the rest of the dining room fittings, was in all probability considered to be "Renaissance" in its day. The pierced filigree work of the base and the small-scaled and delicate but complex ornament of the branches and upper half of this chandelier are typical of the best metal workmanship executed during the 1890s. The six gas candles are interspersed among six upright and six inverted electric lights shaded by delicately etched glass. It is possible that this fixture was a German import. By the turn of the century most, if not all, of the chandeliers of this high quality and elaboration of design were entirely electrified. Gas combination fixtures after 1900 were mostly made in the medium and low-priced range, and after 1910 very few gas or combination gas and electric fixtures were made in any form, cheap or expensive. By 1920, interior gaslighting was, for all practical purposes, a thing of the past. It should be noted, however, that many local ordinances required a few gas burners for emergency lights in places of public assembly until the invention of safe battery-operated lights that could be activated automatically in case of power failures.

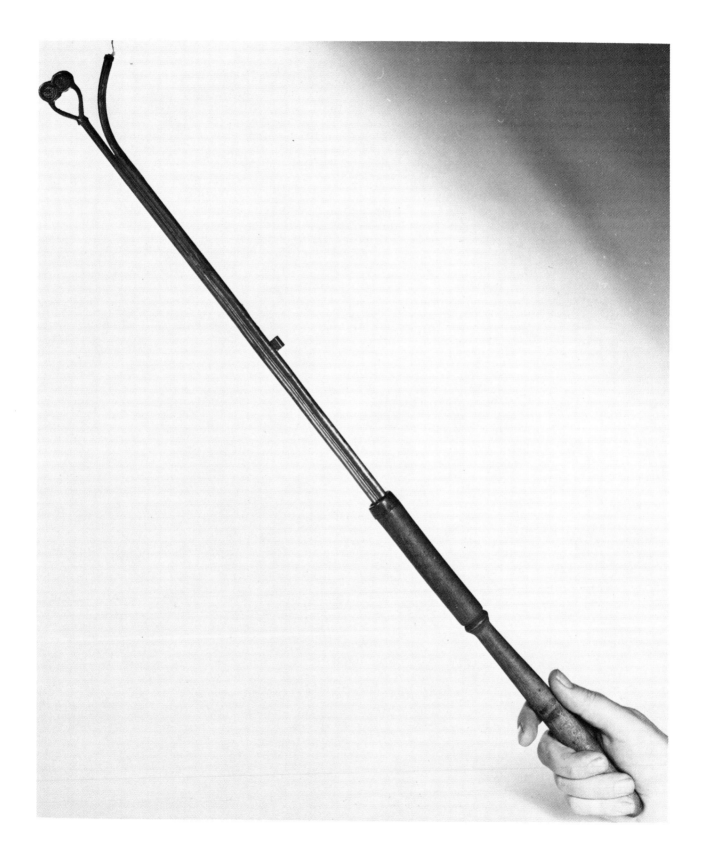

Device for lighting the gas.

Plate **110**

Obviously, gas must be lighted before it can illuminate. While it was easy enough to light a gas wall bracket fixture with a match, chandelier burners were not quite so conveniently accessible. Almost every household using gaslight had what the McKenney and Waterbury catalogue described as a "torch and key lighter," an implement like the one shown here. The "key" was a bifurcated flange designed to grasp a gas key and turn it on or off. The "torch" was a wax taper that could be slid up and down within a metal tube. It worked precisely the way the taper lighters used to ignite altar candles in churches work today. As most chandeliers were hung no more than 7 feet off the floor, almost all lighting and extinguishing of gas burners was done by using lighters like this one. Even chandeliers in places with very high ceilings could be lighted or put out with long-handled torch and key lighters.

Very large and inaccessible fixtures like the huge multijetted chandeliers that hung in some theater auditoria presented special problems. Some had water seal stems, or "pillars," and could be raised and lowered by windlasses like the over 6-ton bronze gas chandelier of 1875 in the Paris Opera. Others, probably because of the difficulty of counterweighting very heavy fixtures, were fixed in position. Therefore, various attempts to light such chandeliers with electric sparks were made. On December 22, 1857, a patent was issued to S. Gardiner, Jr. for a device worked by magnets for "turning on or shutting off inflammable gas by degrees, or gradually, through the agency of electricity," in other words, a gas switch. There is no evidence that it was ever widely adopted for use. On March 30, 1858, the same Mr. Gardiner patented "placing a coil of platinum wire, or its equivalent, in the relative position to the jet or gas described, for the purpose of lighting the jet by electricity, and for re-igniting it when blown out . . ." The current was supplied by a galvanic battery. Gardiner demonstrated his device at the U. S. Capitol and "found that 1,500 burners in the U. S. Senate Chamber at Washington required three seconds to light, including turning on the gas."[159] The inventor's triumph was evidently short-lived, however, as the lighting installation by Cornelius and Baker above the inner skylight of the Senate Chamber, like its counterpart in the House of Representatives, was henceforth lighted "from a small perpetual burner," or pilot light.[160]

Archilaus Wilson devised an igniting system using a Ruhmkorff coil instead of a galvanic battery. He reported on his patent gaslighter before the Franklin Institute in Philadelphia on March 15, 1860, claiming to have overcome such difficulties as the decreasing efficiency of voltaic batteries, fusion of wires, encrustation of wires by soot, cooling of wires by draughts, and failure through the breakage of wires. His list of problems, overcome or not, suggests the flaws common to most electric spark lighting devices of the time. Wilson stated that his invention had lighted a 56-light chandelier "several times with complete success."[161]

In 1861 the Franklin Institute reported yet another gaslighting device as follows:

> Mr. Meyers, of Messrs. Mitchell, Vance and Company, New York, exhibited a neat sample of an apparatus for lighting gas by electricity. The machine consists of a small glass disk, which revolves between two pads of leather, and gives the generated electricity to points, which are in communication with a brass rod about 12 inches long terminating in a ball. An insulated handle is attached to the lower part of the instrument. A piece of wire, attached to a sheath which slips over the burner, is so adjusted that a spark given to it from the ball of the gas lighter passes through the jet of flowing gas and instantly inflames it.[162]

Whether Meyers's gadget was any more successful than Gardiner's or Wilson's has not been ascertained. It is clear, however, that the electric spark method, applied in some way or other, was used after about 1860 to light large and hard to reach chandeliers.

THE WORKS OF THE PEOPLES GAS COMPANY OF BALTIMORE.

Lithograph of the Peoples Gas Company, Baltimore, 1870-1881.

Plate 111

The Peoples Gas Company of Baltimore began operations in 1870, the year it was lithographed in color by A. Hoen for the advertisement reproduced here. This complex was closed down in 1881. Located in southwest Baltimore, the plant was served by the Baltimore and Ohio Railroad, whose coal cars were shunted to the long retort house at the right, where gas was made by reducing the coal to coke. The gas was then piped to be purified by lime in the building at the left, after which it was piped to be distributed from the circular gasholder, or "gasometer," at the far left.[163]

This medium-sized gas plant is typical in appearance of many built throughout the United States during the middle of the 19th century. The long, low brick structures with their monitor roofs resembled those of many other plants, but the immense tank that rose and fell within its tall columnar supports was, of course, a feature unique to gas plants. The tank was constructed of riveted iron plates and was immersed in water at its base. Later gasholders abandoned pseudo-classical columnar supports in favor of plainer ones. A few gasholders were housed in circular brick structures primarily as a precaution against extreme cold.[164]

Many American gashouse neighborhoods looked rather like this one in Baltimore. What the Hoen lithograph does not convey is the sickening stench that made such neighborhoods so noisome. Before lime was given up as a purifying agent in favor of "washers" and "scrubbers" it became extremely foul when saturated with impurities from the gas. It was the used lime rather than the sulfurous odor of the gas itself that made gashouse districts so unpleasant. Such neighborhoods were therefore sometimes inhabited by the least savory elements of the population, tough habitual brawlers and unskilled but nonetheless dangerous criminals.[165]

Although a few isolated instances of gaslighting had occured in America very early in the century, it was not until 1817 that the first company for the manufacture and distribution of illuminating gas was chartered here.[166] The honor of that priority belongs to Baltimore. It has often been stated that the use of gas for lighting did not become widespread in the United States until after the Civil War. That is simply not true. By 1840 there were 11 gaslight companies already chartered, and by 1850 the number had risen to 51. The next decade saw a phenomenal advance, for by 1860 there were 362 companies in the country.[167]

By January 1, 1862, 420 chartered gaslight companies were listed in a compilation made by John B. Murray, a New York dealer in gaslight shares. One of the companies was reported to be making wood gas, three were making water gas and 30 were making rosin gas. All the rest, 386 out of 420 companies, were manufacturing coal gas.[168] As of June 15, 1863, Murray listed 433 chartered companies in this country and 23 in the British Provinces of North America soon to be joined together by the Canadian Confederation of 1867. Aurora, Indiana, was reported to have changed from "Sanders' Water-Gas" to coal gas; and Cold Spring, New York, had changed from rosin gas to coal gas. Scranton, Pennsylvania, was about to try water gas; and the companies at Fishkill, New York, and Smyrna, Delaware, were making petroleum gas. Most significant in the light of future developments was the note that one company, at Freedonia, New York, drew upon an "unlimited supply of natural gas." The gas works at Hoboken, New Jersey, were "not yet built;" and the company at Kittanning, Pennsylvania, was "said to be a fraud." Murray remarked that "Jeff. Davis has extinguished the gas-light of Richmond, Virginia;" and of Williamsburg, Virginia he said, "Camp-fires of the Union Army light this place at present."[169]

It will readily be seen from the statistics cited above that most of the gas manufactured was coal gas and that its use was certainly extensive before the Civil War. The geographical distribution of the chartered companies shows, however, that gas was little used in some areas and heavily used in others. The 433 companies listed by Murray on June 15, 1863, were distributed as follows: 4 in Alabama, 1 in Arkansas, 9 in California, 15 in Connecticut, 4 in Delaware, 1 in the District of Columbia, 1 in Florida, 6 in Georgia, 13 in Illinois, 10 in Indiana, 6 in Iowa, 1 in Kansas, 6 in Kentucky, 2 in Louisiana, 13 in Maine, 7 in Maryland, 58 in Massachusetts, 10 in Michigan, 1 in Minnesota, 4 in Mississippi, 4 in Missouri, 10 in New Hampshire, 21 in New Jersey, 84 in New York, 8 in North Carolina, 35 in Ohio, 1 in Oregon, 53 in Pennsylvania, 7 in Rhode Island, 3 in South Carolina, 4 in Tennessee, 3 in Texas, 8 in Vermont, 10 in Virginia, 1 in West Virginia and 9 in Wisconsin.[170] Clearly, the heavily industrialized states had the most gas plants, with Massachusetts, New Jersey, New York, Ohio, and Pennsylvania far in the lead. As one would expect, the thickly settled areas had the gas plants and the rural districts depended upon lamps and candles or, for very grand establishments, gas machines (see Appendix).

From the Commercial Credit Company Collection, Baltimore, photograph courtesy of the Maryland Historical Society, Baltimore, Maryland.

Ornamented gasmeter, ca. 1867.

Plate **112**

The first gasmeters known to have been made in the United States were manufactured by Samuel Hill in Baltimore about 1832. John Rodger, a Baltimore machinist working under the supervision of John M. Slaney, made the second meter known, probably in the year 1832. The third American gasmeter maker is said to have been one C. Young of New York, who produced his meter around 1835. The fourth maker was the firm of Colton and Code around 1839 in Philadelphia.[171] On October 7, 1834, James Bogardus of New York obtained a patent for a gasmeter. It was made of cast iron, and "its exterior form is that of an oval shade, such as are put over time pieces, the longest diameter of which is about nine inches, its shortest six or seven, and its height fourteen."[162] Bogardus's meter, from its description, appears to have been a dry gasmeter operating on a bellows, or diaphragm principle, although dry meters are commonly said to have supplanted wet meters only in 1844 or thereafter. James Bogardus (1800-1874) is best known for his important role in the development of cast-iron buildings in America.

Albert Potts of Philadelphia patented a meter on July 5, 1859, for which new convenience was claimed.

> The advantage of this improvement consists in the facility with which the Meters may, at any time, be inspected by the Agents of the Gas Company without necessarily being obliged to enter the premises . . . the meter is nicely adjusted to an auxiliary case which is imbedded in the Front Wall of the Building. . . . With this improvement the Meter is less liable to Freeze in Winter season, as the inside portion of the Receptacle is exposed to the heat from the Parlor or Room. . . .

Evidently Pott's scheme was not very widely adopted, as most gasmeters remained housed entirely indoors. The gasmeters in the U.S. Treasury Department Building in Washington, D.C., were kept from freezing by the addition of alcoholic spirits, four gallons of which were requisitioned for this purpose in 1852.[173]

Most gasmeters were plain and strictly ulititarian objects and were placed in basements, cellars, or service areas of buildings. Occasionally, however, meters were ornamented and placed in parlors. The example illustrated here is painted a bright red and decorated with painted putti and roses stencilled in gilt. It was manufactured by William Wallace Goodwin (1833-1901), a Philadelphia maker who founded W. W. Goodwin and Company about 1872.[174] Goodwin was granted a dry gasmeter patent (no. 76,908) on April 21, 1868, and that patent date was subsequently antedated to November 5, 1867.[175] The patent was for the mechanism, not for the ornamentation. This extraordinary object was found on a decorative shelf in the parlor of a house on Walnut Street in Philadelphia. It is analogous, in a way, to other fancily painted mechanical devices of its period. Americans took great pride in their ingenious gadgets, and just about every new invention, from sewing machines to railroad locomotives, had ornamentation lavished upon it.

From the private collection of James M. Goode, Washington, D.C. Smithsonian Institution photograph.

233

Exterior lighting in general and street lighting by gas in particular have not been considered in the preceding text and illustrations, because they constitute a separate subdivision of the principal subject under consideration, i.e., the lighting of buildings by gas. Essentially, differences of practicality, purpose, and use divided exterior from interior gaslighting.

Until after 1890, the usual gas street lamp in American cities and towns was similar to the example shown here, a cast-iron standard of comparatively simple design surmounted by a square glass lantern enclosing a batswing burner. Frequently, as on this Washington, D. C., lamppost, seen in a detail from an 1865 photograph by one of Matthew Brady's staff named Smith, there was a crossbar below the lantern against which the lamplighter's ladder was steadied. Lampposts were made of iron cast in a variety of plain and fancy patterns. Usually, they were standarized in various neighborhoods, as evident from old photographic city or town views such as this, which provide valid information respecting the design of the lampposts formerly used in specific locales. Fortunately, modern recastings from molds taken off old fixtures are, for all practical purposes, indistinquishable from old castings. (When researching for a restoration, be sure that the lampposts in an old view were in use at the time to be represented.)

Occasionally, the cost of a post was saved by running a pipe up the side of a building and cantilevering a lantern out on a bracket attached to the building's corner. Corner lamps, whether on posts or attached to buildings, often had translucent glass sections set in the upper rims of the lanterns on which street names were lettered. Most major iron foundries made street lamps. Morris, Tasker and Company of Philadelphia made lampposts that differed only in minor details from this Washington example.[176] The most noticeable difference between this fixture and a number of Philadelphia examples was that the latter had finials in the form of eagles atop their lanterns.

The limited amount of light that was normal until the introduction of the Welsbach burner would not be tolerated within the limits of any American municipality today. Considerations of safety alone preclude that. In a number of "gas-light districts" that have recently been established for "atmosphere," it has been the invariable practice to provide more lamp standards (almost always with mantles instead of flat-flame burners) than would ever have been the case on even the best-lighted 19th century streets, often with auxiliary electric street lighting as well. In attempting to recreate early gas street lighting, major compromises are essential. It is probably only under museum conditions that really authentic gaslighting of the kind in use before ca. 1890 can be installed.

Unfortunately, many modern gas street lamps are not provided by suppliers, especially gas companies, with any means for turning them off. A street lamp left burning during the light hours of the gas era would have been an anomaly indeed!

From the Library of Congress, the Brady Collection.

Philadelphia theater street lamps, 1860.

Plate 114

Theaters, saloons, dance halls and other places of evening public entertainment frequently supplemented the feeble municipal street lighting with lighting of their own that served for advertising as well as illumination of the exterior stairs and curb. This photograph of the Arch Street Theatre in Philadelphia, taken by John McAllister in 1860, shows two large-scaled gas lanterns supplied by the management as well as a standard Philadelphia street lamp with eagle finial. The two theater lanterns are typical of those used in front of many places of public resort. Note the difference in scale between the lighting fixtures provided by the theater and that provided by the city.

A single large gas lantern once stood before the central entrance of Ford's Theatre in Washington, D. C. When the theater was restored by the National Park Service and reopened in 1968, a new gas lantern, scaled and proportioned according to the evidence contained in old photographs, was erected to replace the long-vanished original. Shown here during installation on the post, the new lantern reproduces the original one and is slightly over 5 feet in height.

A compromise with strict historical accuracy was made with respect to the burners in order to meet modern lighting requirements. The burners now in use resemble Argand burners without chimneys and were specially developed by the Welsbach Company to give a greater intensity of light than any standard open flame burner. As gas mantles would have been obvious anachronisms, the needed amount of light was obtained by stretching wires across the path of the open flame. The wires are not visible from the street level, yet they become incandescent and thus emit the needed light.[177] The lamppost itself was cast from a mold taken of an original post of the same pattern as that shown in old photographs of Ford's Theatre.

Lamppost for private driveway, ca. 1850.

Plate **116**

Lampposts illuminating private driveways were never very common, but a few examples installed during the last century have been recorded. The example shown here stood until 1937 beside the carriage drive in the grounds of the Valentine-Fuller House, built in 1848 in Cambridge, Massachusetts. The octagonal post is perhaps slightly more elaborate than standard street lampposts in use around 1850. But the lamp itself is decidedly more ornate than the square lanterns used for street lighting. Fancy octagonal lantern forms, when used outdoors, were found only in association with major public buildings or private installations of gaslighting. Note the suggestion of chinoiserie in the pagoda-like form of the lamp's finial.

From the Library of Congress, Historic American Buildings Survey [Mass-283A], photograph by Richard Ruggles.

Welsbach burner street lamps in Independence Square, Philadelphia, 1905.

Plate **117**

Between 1890 and 1900 street lighting was radically improved by the widespread introduction of the Welsbach burner. Simultaneously, the old square lantern type of lamp was replaced by the white-domed cylindrical type seen here. Cylindrical glass was easier to clean than the four sides of a square lantern, and it would not crack in proximity to a flame that was confined within a gas mantle. The white glass dome was also a substantial improvement, as it reflected the light downward, where it was wanted. The new fixtures were normally mounted on already existing posts, as was the case with the light shown here in a detail from a 1905 photograph of the south front of Independence Hall in Philadelphia. The post, which has a complex base composed of alternating square and circular elements of Eastlake detailing and a rather Baroque spirally twisted shaft, appears to date from around 1876. It is much more elaborate than normal street light standards, as it is one of a series designed to light the walkways of Independence Square, the park-like setting of one of America's principal historic structures. After 1900 the type of lamp seen here became ubiquitous in America, and the older lantern form of street lamp all but totally disappeared.[178]

Cast-iron ornamental lamp standards at Tennessee State Capitol, 1859.

Plate **118**

The principal entrances of important public buildings were often flanked by ornamental lamp standards, but certainly few were as elaborate as the pair seen here in a photograph taken during the Civil War by one of Matthew Brady's staff. This pair once lighted the south entrance steps of the Tennessee State Capitol in Nashville, and the building's other three axial entrances had pairs of cast-iron lamp standards identical with these. All were cast by the Philadelphia foundry of Wood and Perrot by 1859, the year the capitol was completed. Their designer is unknown. The figures grouped around the columnar shafts represent morning, noon, and night.[179] Groups of large-sized figures such as these have not been recorded in association with any other light standards in America and were probably unique to the Nashville examples.

Although the cluster of three figures standing on their hexagonal base and surrounding a column of the composite order gives an effect of considerable richness, that effect is diminished by the fact that all the sculptural groups are identical and therefore clearly produced from casting molds rather than from the white marble they are intended to resemble. Note the complex glazing pattern of the lavishly ornamented octagonal lanterns and the figurines that form their finials. As often happens with the changing tastes of the times, those splended lampposts were all discarded during modernization of the Tennessee Capitol.

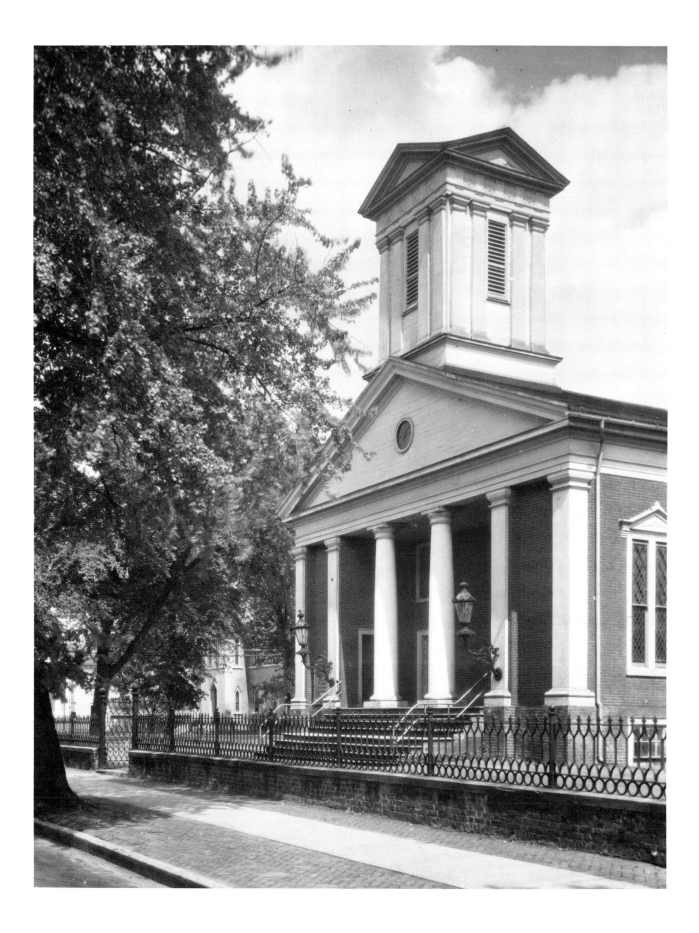

Cast-iron exterior lamp brackets, probably 1854.

Plate **119**

T he cast-iron exterior lamps and brackets flanking the portico of the Presbyterian Church of 1833 in Fredericksburg, Virginia, in this photograph by Frances Benjamin Johnston, may be the only surviving examples of their particular type in America. The local gaslight company in Fredericksburg was chartered in 1854, and this pair of lamps probably dates from that year. Exterior lighting actually attached to buildings (aside from some corner street lamps) was not common, except for standards on plinths flanking entrances. These bracketed lamps in Fredericksburg are further noteworthy because they closely resemble the pairs of lamps that were attached to the outermost columns of the north portico of the White House in Washington from around 1848 until McKim, Mead and White's renovations of 1902. It should be noted, however, that the entrances of the Academy of Music in Philadelphia, completed in 1857, were lighted by similar bracketed lamps, now replaced by modern replicas. The entire facade of the Boston "Museum," a theater built in 1846, was lighted by tiers of gas brackets, a highly exceptional treatment.[180]

There was at least one instance of a shop with extensive exterior illumination. A Duval lithograph illustrating the cover of "The New Costume Polka" with a view of The Lee and Walker's Music Store in Philadelphia was published in 1851 and dedicated to Mrs. Amelia Bloomer, the dress reformer. The rail of the balcony balustrade that runs across the facade is topped by at least seven gaslights with large globular shades.

Indirect exterior lighting was actually in use as early as 1838. The diarist Philip Hone, once Mayor of New York City, writing about architect William Strickland's Second Bank of the United States on February 14, 1838, entered the following passage in his journal:

> The portico of this glorious edifice, the sight of which always repays me for coming to Philadelphia, appeared more beautiful to me this evening than usual, from the effect of the gaslight. Each of the fluted columns had a jet of light from the inner side so placed as not to be seen from the street, but casting a strong light upon the front of the building the softness of which, with its flickering from the wind, produced an effect strikingly beautiful.[181]

Appendix

This chronological listing is based upon the alphabetical listing compiled by John B. Murray, a New York dealer in gaslight shares, and published in *The American Gas-Light Journal*, June 15, 1863, pp. 370-373. The dates given are the years when the gas companies received their charters from their respective state legislatures. Some companies began operations *before* their charters were formally approved; others did not begin operations until a year or more *after* their charters were granted.

1817 Baltimore, Maryland
1822 Boston, Massachusetts
1823 New York, New York
1825 Brooklyn, New York
1830 Manhattan (New York City), New York
1833 Evansville, Indiana
1835 New Orleans, Louisiana
    Pittsburgh, Pennsylvania
1836 Monroe, Michigan
1838 Louisville, Kentucky
1839 St. Louis, Missouri
1841 Cincinnati, Ohio
1844 Philadelphia, Pennsylvania
1845 Albany, New York
1846 Charleston, South Carolina
    Cleveland, Ohio
    Newark, New Jersey
1847 Fall River, Massachusetts
    Milwaukee, Wisconsin
    New Haven, Connecticut
    Paterson, New Jersey
    Trenton, New Jersey
1848 Buffalo, New York
    Dayton, Ohio
    Providence, Rhode Island
    Reading, Pennsylvania
    Springfield, Massachusetts
    Syracuse, New York
    Troy, New York
    Washington, Pennsylvania
    Zanesvile, Ohio
1849 Chicago, Illinois
    Detroit, Michigan
    Great Falls, New Hampshire
    Hartford, Connecticut
    Lancaster, Pennsylvania
    Lawrence, Massachusetts
    Lowell, Massachusetts
    Portland, Maine
    Savannah, Georgia
    Utica, New York
    Worcester, Massachusetts
    York, Pennsylvania
1850 Columbus, Ohio
    Easton, Pennsylvania
    Nashville, Tennessee
    Pawtucket, Rhode Island
    Salem, Massachusetts
    Wheeling, Virginia (West Virginia)
    Williamsburg, New York
    Yreka, California

1851 Alexandria, Virginia
    Augusta, Georgia
    Bridgeport, Connecticut
    Charlestown, Massachusetts
    Chelsea, Massachusetts
    Chillicothe, Ohio
    Frankfort, Kentucky
    Frederick City, Maryland
    Germantown, Pennsylvania
    Madison, Indiana
    Portsmouth, New Hampshire
    Richmond, Virginia
    Schenectady, New York
1852 Allegheny, Pennsylvania
    Bangor, Maine
    Burlington, New Jersey
    Cambridge, Massachusetts
    Camden, New Jersey
    Columbia, Pennsylvania
    Columbia, South Carolina
    Elmira, New York
    Erie, Pennsylvania
    Ithaca, New York
    Lewiston, Maine
    Lynchburg, Virginia
    Lynn, Massachusetts
    Macon, Georgia
    Manayunk, Pennsylvania
    Manchester, New Hampshire
    Memphis, Tennessee
    Mobile, Alabama
    Montgomery, Alabama
    Nashua, New Hampshire
    New Albany, Indiana
    New Bedford, Massachusetts
    Newburg, New York
    Newburyport, Massachusetts
    Oswego, New York
    Petersburg, Virginia
    Rochester, New York
    Rome, New York
    San Francisco, California
    Waterford, New York
    Watertown, New York
    West Chester, Pennsylvania
    Wilmington, Delaware
1853 Allentown, Pennsylvania
    Augusta and Hallowell, Maine
    Bath, Maine
    Binghampton, New York
    Bordentown, New Jersey

Brookline, Massachusetts
Burlington, Vermont
Canandaigua, New York
Cape Island, New Jersey
Columbus, Georgia
Concord, New Hampshire
Covington and Newport, Kentucky
Dedham, Massachusetts
East Boston, Massachusetts
Fitchburg, Massachusetts
Geneva, New York
Greenpoint, New York
Haverhill, Massachusetts
Hudson, New York
Jamaica Plain, Massachusetts
Jersey City, New Jersey
Lexington, Kentucky
Little Falls, New York
Middletown, Connecticut
New London, Connecticut
Newport, Rhode Island
Norristown, Pennsylvania
Plymouth, Massachusetts
Quincy, Illinois
Saratoga Springs, New York
Taunton, Massachusetts
Toledo, Ohio
Vicksburg, Mississippi
Waltham, Massachusetts
West Troy, New York
Wilmington, North Carolina
Woonsocket, Rhode Island

1854 Bethlehem, Pennsylvania
Cohoes, New York
Cumberland, Maryland
Dorchester, Massachusetts
Dubuque, Iowa
Fredericksburg, Virginia
Gardiner, Maine
Glenn's Falls, New York
Gloucester, Massachusetts
Hagerstown, Maryland
Honesdale, Pennsylvania
Knoxville, Tennessee
Lafayette, Indiana
Malden and Melrose, Massachusetts
Marblehead, Massachusetts
Mount Holly, New Jersey
Nantucket, Massachusetts
New Brunswick, New Jersey
Newton and Watertown, Massachusetts
Norwalk, Connecticut
Norwich, Connecticut
Pittsfield, Massachusetts
Rock Island, Illinois
Rockland, Maine
Rondout and Kingston, New York
Roxbury, Massachusetts

Sandusky, Ohio
Springfield, Illinois
Warren, Rhode Island
Waterbury, Connecticut
West Cambridge, Massachusetts
Winchester, Virginia
Yonkers, New York

1855 Atlanta, Georgia
Batavia, New York
Bloomington, Illinois
Bristol, Rhode Island
Carlisle, Pennsylvania
Davenport, Iowa
Elizabeth, New Jersey
Flushing, New York
Galena, Illinois
Great Barrington, Massachusetts
Harrisburg, Pennsylvania
Keokuk, Iowa
Kingston, Pennsylvania
Lansingburg, New York
Madison, Wisconsin
Meadville, Pennsylvania
North Attleboro, Massachusetts
Ogdensburg, New York
Ottawa, Illinois
Peekskill, New York
Peoria, Illinois
Piqua, Ohio
Portsmouth, Ohio
Portsmouth, Virginia
Racine, Wisconsin
Stamford, Connecticut
Washington, D.C.
Xenia, Ohio

1856 Akron, Ohio
Bellefont, Pennsylvania
Bristol, Pennsylvania
Canton, Ohio
Catasequa, Pennsylvania
Chester, Pennsylvania
Danville, Pennsylvania
Eastport, Maine
Freeport, Illinois
Hollidaysburg, Pennsylvania
Huntsville, Alabama
Jamaica, New York
Janesville, Wisconsin
Johnstown, Pennsylvania
Lambertville, New Jersey
Lancaster, Ohio
Lebanon, Pennsylvania
Lewistown, Pennsylvania
Montpelier, Vermont
Newark, Ohio
Northampton, Massachusetts
Norwalk, Connecticut
Owego, New York

Paducah, Kentucky
Palmyra, New York
Pottstown, Pennsylvania
Pottsville, Pennsylvania
Poughkeepsie, New York
Salem, New Jersey
Seneca Falls and Waterloo, New York
Sing Sing, New York
Urbana, Ohio
Watertown, Wisconsin
Wilkesbarre, Pennsylvania
Woodstock, Vermont
Wooster, Ohio
1857 Adrian, Michigan
Alton, Illinois
Attleboro, Massachusetts
Bridgeton, New Jersey
Chambersburg, Pennsylvania
Chicopee, Massachusetts
Columbia, California
Danbury, Connecticut
Doylestown, Pennsylvania
Easton, Maryland
Fishkill, New York
Gloversville, New York
Harlem, New York
Huntington, Pennsylvania
Iowa City, Iowa
Jackson, Michigan
Joliet, Illinois
Marietta, Ohio
Massilon, Ohio
Muscatine, Iowa
Newcastle, Delaware
Rahway, New Jersey
Rockford, Illinois
Sacramento, California
Saint Joseph, Missouri
Saint Paul, Minnesota
Selma, Alabama
Staten Island, New York
Ware, Massachusetts
Waverly, Mississippi
West Point, New York
Williamsport, Pennsylvania
1858 Albion, New York
Astoria, New York
Belfast, Maine
Catskill, New York
Charlotte, North Carolina
Clinton, Massachusetts
East Greenwich, Rhode Island
Fort Wayne, Indiana
Freedonia, New York (natural gas)
Fulton, New York
Galveston, Texas
Grand Rapids, Michigan
Greensburg, Pennsylvania

Jacksonville, Missouri
Jersey Shore, Pennsylvania
Kalamazoo, Michigan
Lewisburg, Pennsylvania
Lockhaven, Pennsylvania
Marysville, California
Natchez, Mississippi
Placerville, California
Raleigh, North Carolina
Saint Albans, Vermont
Scranton Gas and Water, Pennsylvania
Walden, New York
Wautoma, Wisconsin
Williamsburg, Virginia
Ypsilanti, Michigan
1859 Altoona, Pennsylvania
Amesbury and Salisbury, Massachusetts
Auburn, New York
Bath, New York
Beverly, Massachusetts
Birmingham, Pennsylvania
Brattleboro, Vermont
Brockport, New York
Burlington, Iowa
Charlotteville, Vermont
Citizens', Brooklyn, New York
Clyde, New York
Dover, Delaware
Fayetteville, North Carolina
Flemington, New Jersey
Freemont, Ohio
Galion, Ohio
Gettysburg, Pennsylvania
Greensboro, North Carolina
Hannibal, Missouri
Haverstraw, New York
Hempstead, New York
Kittanning, Pennsylvania (questionable)
La Crosse City, Wisconsin
Leavenworth, Kansas
Lyons, New York
Metropolitan (New York City), New York
Middleboro, Massachusetts
Morrisania, New York
Morristown, New Jersey
New Bern, North Carolina
Niagara Falls, New York
North Bridgewater, Massachusetts
Norwalk, Ohio
Nyack, New York
Orange, New Jersey
Peoples', Chicago, Illinois
Plattsburg, New York
Portland, Oregon
Port Lavacca, Texas
Richmond, Massachusetts
Ripley, Ohio
Rutland, Vermont

251

Sag Harbor, New York
Salem, Ohio
Salisbury, North Carolina
Sandwich, Massachusetts
Southbridge, Massachusetts
Staunton, Virginia
Tarrytown and Irvington, New York
Vincennes, Indiana
Washington, North Carolina
Willimantic, Connecticut
1860 Aurora, Indiana
Carbondale, Pennsylvania
Claremont, New Hampshire
Cold Spring, New York
Cold Water, Michigan
Delaware, Ohio
Dobbs' Ferry, New York
Greenfield, Massachusetts
Hornellsville, New York
Keene, New Hampshire
Laconia, New Hampshire

Oberlin, Ohio
Owensboro, Kentucky
Penn Yan, New York
Plainfield, New Jersey
Port Chester, New York
Port Jervis, New York
South Adams, Massachusetts
1861 Battle Creek, Michigan
Calais, Maine
East Hampton, Massachusetts
East New York, New York
Hayesville, Massachusetts
Holyoke, Massachusetts
Homer and Cortlandt, New York
Le Roy, New York
Little Rock, Arkansas
Norwich, New York
Painesville, Ohio
Warren, Ohio
1862 Hoboken, New Jersey (works not built as of June 15, 1863)

Gas Companies listed on June 15, 1863, for which no charter date is given

Beloit, Wisconsin
Brownville, Pennsylvania
Clarkesville, Tennessee
Dover, New Hampshire
Ellicotts' Mills, Maryland
Exeter, New Hampshire
Fair Haven, Connecticut
Fifth Ward (Milwaukee), Wisconsin
Hamilton, Ohio
Hastings, New York
Indianapolis, Indiana
Jackson, Mississippi
Jackson, California
Jacksonville, Florida
Jacksonville, Indiana
Jamestown, New York
Jeffersonville, Maryland
Lockport, New York
Mansfield, Ohio
Marlboro, Massachusetts
Mauch Chunk, Pennsylvania
Medina, New York
Meriden, Connecticut
Middletown, New York
Milford, Massachusetts
Milledgeville, Georgia
Milton, Pennsylvania
Mount Vernon, Ohio
New Britain, Connecticut

Newton, New Jersey
Newton, Pennsylvania
Norfolk, Virginia
Northern Liberties, Pennsylvania
North Adams, Massachusetts
Quincy, Massachusetts
Saco, Maine
Saint Johnsbury, Vermont
Salisbury Mills, Massachusetts
Salmon Falls, Maine
San Antonio, Texas
San Diego, California
Saugerties, New York
Shreveport, Louisiana
Smyrna, Delaware
South Boston, Massachusetts
Springfield, Ohio
Steubenville, Ohio
Stockton, California
Tamaqua, Pennsylvania
Terre Haute, Indiana
Thompsonville, Connecticut
Tiffin, Ohio
Westfield, Massachusetts
Weston, Ohio
White Plains, New York
Woburn, Massachusetts
Yorkville, South Carolina

Notes

1. Fredrick Accum, *A Practical Treatise on Gas-Light* (London: R. Ackermann, 1815), p. 115. Accum's name, given as Fredrick on the title page of his book, was actually Friedrich Christian Accum.

2. This remarkably early view of a gaslighted interior has been reproduced in color in J. B. Priestley, *The Prince of Pleasure and his Regency 1811-1820* (New York: Harper & Row, 1969), plate 15.

3. Accum, *Practical Treatise*, p. 118. Accum is actually referring here to his plate 4, but his statement applies with equal validity to his plate 5. His description of his plates 3-5 is found on pp. 115-121.

4. Dean Chandler, *Outline of History of Lighting by Gas* (London: South Metropolitan Gas Company, 1936), p. 200. This dates the first advertisement for the Welsbach mantle as having appeared in December 1890.

5. *Ibid.*, p. 87.

6. Letter of August 20, 1976, from J. P. A. Scott to W. Brown Morton III, in Office of Archeology and Historic Preservation files. Scott, who is Museum Assistant, Illumination Collection, at the Science Museum, South Kensington, London, dates the engravings between 1820 and 1830 on the basis of firmly dated fixtures in the museum's collection.

7. *Journal of the Franklin Institute*, 3rd ser. (Philadelphia: Franklin Institute), 37(1859): 125-126; an advertisement in *The American Gas-Light Journal*, 2(1860):164 gives the American patent date.

8. Starr, Fellows, and Company, *Illustrated Catalogue of Lamps, Gas Fixtures, &c.* ([New York], 1856), plates 31, 41. The Academy of Music in New York, built in 1854, had brackets in the form of chimerical creatures (plate 29) and the hall chandelier of 1854 in the Wickham-Valentine House (Valentine Museum) in Richmond, Virginia, has branches in the form of dragons. The January 1973 issue of *Antiques* shows a full-color view of the Richmond example. The rather repulsive (except for a herpetologist!) bracket in the form of a rattlesnake was one of a series of fixtures designed in 1859 by Joseph Goldsborough Bruff (1804-1889) for the south wing of the U.S. Department of the Treasury in Washington; Washington, D. C., National Archives, Record Group 121, Records of the Public Buildings Service, "Photographs of Designs, Bureau of Construction, Treasury Department."

9. First National Bank in St. Louis, *St. Louis—A Fond Look Back* (St. Louis, 1956). A lithograph of J. Y. Hart's "Capitol Oyster Saloon and Restaurant" shows counterweighted water-seal chandeliers with smoke bells over the burners. The Seabury Tredwell House (Old Merchant's House) of 1832 in New York City has a fine bronze-finished pair of neoclassical counterweighted water-seal gaseliers that appear to be as old as the house. Curiously, the James Lancaster Morgan House of 1870 in Brooklyn had a counterweighted water-seal gaselier in the front hall as well as one in the dining room that exemplified the Neo-Grec style; John A. Kouwenhoven, *The Columbia Historical Portrait of New York* (Garden City: Doubleday and Company, 1953), pp. 369-370; the Morgan dining room is also illustrated in William Seale, *The Tasteful Interlude* (New York: Praeger Publishers, 1975), p. 103.

10. Stereograph by G. W. Wilson, Aberdeen, "Abbotsford—The Library. No. 574," author's collection.

11. Edward Everett Hale, *A New England Boyhood* (Boston: Little, Brown and Company, 1927), p. 5. Hale, born in 1822 and writing in 1893, was reporting a personally observed circumstance.

12. Marjorie Drake Ross, *The Book of Boston—The Victorian Period—1837 to 1901* (New York: Hastings House Publishers, 1964), p. 71.

13. Marshall B. Davidson, *Life in America* (Boston: Houghton Mifflin Company, 1951), 2:12. The scene shown in the Thayer lithograph is identified as "Tremont House;" and as "the ballroom of Tremont House" in Marshall B. Davidson, *The American Heritage History of American Antiques from the Revolution to the Civil War* (New York: American Heritage

Publishing Company, 1968), p. 331. However, William Havard Eliot, *A Description of Tremont House with Architectural Illustrations* (Boston: Gray and Bowen, 1830), shows that no such ballroom existed in the hotel.

14. Edgar Allan Poe, "Philosophy of Furniture" (1840), *The Complete Tales and Poems of Edgar Allan Poe* (New York: The Modern Library, 1938), p. 464.

15. Voucher, "Miscellaneous Treasury Accounts of the General Accounting Office, 1790-1894," National Archives, Record Group 130. Information courtesy of William Seale. Loris S. Russell, "Early 19th Century Lighting," *Building Early America* (Radnor, Pennsylvania: Chilton Book Company, 1976), p. 197. Russell states the chandeliers were "oil fixtures;" he was misinformed.

16. Oliver E. Allen, "The Lewis Albums," *American Heritage* (December 1962), pp. 65-80; includes color illustration.

17. "The cast pillar icicles, and other pendulous ornaments of these splendid lamps, are the first of the kind presented to the public from American sources, and they bear a strict scrutiny for transparency, lustre, and workmanship." *Journal of the Franklin Institute*, n.s. 13 (Philadelphia, 1834), p. 93.

18. A handsome G. and W. Endicott lithograph in the Bella C. Landauer Collection at the New York Historical Society shows three chandeliers and two brackets in the ladies' saloon. The ill-fated "Atlantic" made her first trip on August 18, 1846, and was lost by shipwreck in November of that same year. She manufactured her gas on board. John H. Morrison, *History of American Steam Navigation*, (New York: Stephen Daye Press, 1958), p. 328. Except for Long Island Sound and Hudson River vessels, few steamboats used gaslighting. The flexing of limber-hulled Western river boats made gas impracticable, as did the motion of ocean-going ships. Hudson River steamboats used compressed, or "liquid," gas to supply their chandeliers. *Doggett's New York City Directory* for 1846-1847 listed "Charles Starr, liquid gas," at 117 Fulton Street. *Wilson's Business Directory of New York City* for 1860 listed six firms under the heading of "Gas Works (Portable)," among them the "New York Car and Steamboat Company" at the premises listed in 1846-1847 for Starr. On October 1, 1859, *The American Gas-Light Journal* carried an advertisement by S. B. Bowles for "gas apparatus for cars."

19. Cf., plate 50 of this report. After Charles Goodyear patented his vulcanization process on June 15, 1844, it became possible to make rubber hose with which to attach gaslamps to chandeliers, pendants, or brackets. A lithograph illustration in Charles Ellery Stedman [Chinks], *Mr. Hardy Lee, His Yacht* (Boston: A. Williams and Company, 1857) shows a lamp suspended from a chandelier by what appears to be a flexible hose, although it may be a very slender metal duct. The lamp has an apparently paper octagonal shade, and jointed brackets in another illustration in Stedman's pseudonymous work have what seem to be fluted paper shades. These and other lithographs by "Chinks" are reproduced in *American Heritage* (June 1964), pp. 24-31.

20. Thomas U [stick] Walter and J. Jay Smith, *A Guide to Workers in Metal and Stone: For the Use of Architects and Designers, Black and White Smiths, Brass Founders, Gas Fitters, Iron Masters, Plumbers, Silver and Goldsmiths, Stove and Furnace Manufacturers, Pattern Makers, Marble Masons, Stucco Workers, Carvers and Ornamental Workers in Wood, Potters, Etc., From Original Designs, and Selections Made from Every Accessible Source, American and European* (Philadelphia: Carey and Hart, 1846), plate 17.

21. The George Washington Whittemore House of 1850 that formerly stood at 329 Harvard Street in Cambridge, Massachusetts, had four brackets, each with seven gas candles, in the parlor. The sleeves forming the "candles" were porcelain.

22. "Cornelius & Baker are the most extensive manufacturers of lamps, chandeliers, gas-fixtures, &c., in the United States, employing upwards of seven hundred persons in the several departments of the establishment. . . ." *The Art-Journal Illustrated Catalogue — The Industry of All Nations 1851* (London: George Virtue, 1851), p. 212; "The pioneer establishment in this manufacture, [of lamps and chandeliers] and the one which, in extent,

is now confessedly without an equal in Europe or America, is that of Cornelius & Baker." Edwin T. Freedley, *Philadelphia and Its Manufactures: A Hand-Book Exhibiting the Development, Variety, and Statistics of the Manufacturing Industry of Philadelphia in 1857. Together with Sketches of Remarkable Manufactories; and a List of Articles Now Made in Philadelphia* (Philadelphia: Edward Young, 1859), p. 352; "To-day a single firm or establishment in this country, that of the Messrs. Cornelius and Sons, of Philadelphia, Penn., makes nearly one-half of all the gas fixtures manufactured in the United States, which, together with the unsurpassed, if not wholly unequalled character as well, of their wares, renders them the representative manufacturers in their line." Horace Greeley, *et al, The Great Industries of the United States: Being an Historical Summary of the Origin, Growth, and Perfection of the Chief Industrial Arts of this Country* (Hartford: J. B. Burr and Hyde, 1873), p. 308. From these quotations it would appear that the Cornelius firm held its lead from before 1851 until at least 1873.

23. Charles S. Cornelius, *History of the Cornelius Family in America* (Grand Rapids, Michigan, 1926), 8, pp. 2-4; 10, pp. 48-49.

24. Pyne Press, *Lamps & Other Lighting Devices — 1850-1906* (Princeton: The Pyne Press, 1972), p. 21. Instead of 1783, "about 1800" is also mentioned as the time of Christian Cornelius's arrival. Greeley, *Great Industries*, p. 315.

25. Pyne Press, *Lamps*, p. 21.

26. Dean Hale, "Diary of an Industry," *The American Gas Journal* (1970), unpaginated reprint.

27. Greeley, *Great Industries*, p. 315.

28. *Ibid.* "In 1831, Robert was admitted to the partnership, under the style Cornelius and Son. . . ." However, a letter dated June 30, 1840, to Colonel S. Birdsall from the firm, stating that fixtures for the North Carolina Capitol at Raleigh had been shipped, was signed "Cornelius & Co." North Carolina Archives, ST P&C of SC 1839-45. It should be noted that those fixtures were fitted for real candles, not gas. The print in Walter's *Guide to Workers in Metal and Stone* refers to "Cornelius and Son," but *McElroy's Philadelphia Directory* consistently lists the firm as "Cornelius & Co." from 1841 through 1856. As has already been seen in footnote 22, the firm was styled "Cornelius & Baker" by the *Art-Journal* as early as 1851. It therefore seems reasonable, despite directory listings, to date the change from "Cornelius & Co." (or "Cornelius & Son") to "Cornelius, Baker & Co." from the death of Christian Cornelius in 1851.

29. *Directory of American Biography*, 1935 ed., s.v. "Gerard Troost." Dr. Gerard Troost had been sent on a scientific expedition to Java by King Louis of Holland before coming to Philadelphia in 1810. He was a founder and first President of the Academy of Natural Sciences in Philadelphia and went in Robert Dale Owen's famous "boat-load of knowledge" to New Harmony, Indiana in 1825. In 1827 he went to Nashville, Tennessee, where he joined the faculty of the University of Nashville in 1828. He was one of the most eminent American scientists of his day.
George C. Groce and David H. Wallace, *The New York Historical Society's Dictionary of Artists in America — 1564-1860* (New Haven: Yale University Press, 1957), pp. 151-152. James Cox was an English born-Philadelphia artist of considerable prominence as a teacher. He had a library of over 5,000 books on art, a very large collection for the time, which he ultimately sold to the Library Company of Philadelphia. If one may judge from the caliber of his teachers, Robert Cornelius could hardly have received better instruction.

30. Greeley, *Great Industries*, p. 315.
This remarkably early daguerreotype portrait has been reproduced in *American Heritage* (December 1956), p. 50; an excellent account of Robert Cornelius' career as a daguerreotypist is contained in Philadelphia Museum of Art, *Philadelphia, Three Centuries of American Art, Bicentennial Exhibition, April 11 to October 10, 1976* (Philadelphia, 1976), pp. 311-313.

31. Dean Hale, "Diary of an Industry," *The American Gas Journal* (1970), unpaginated reprint.

32. Reproduced in *American Heritage* (October 1969), p. 48. The ultimate source is not noted,

but was probably *Harper's Weekly* or *Frank Leslie's Illustrated Newspaper.*

33. Photographs taken in 1916, now in the Armory Museum files, and an old photograph of a Myers House interior reproduced in William Rotch Ware, *The Georgian Period,* Boston, American Architect and Building News Company, 1899-1902, illustrate these chandeliers.

34. The inversion of the detail in the Baltimore examples may have been the result of the original assembly, but it suggests that it is highly advisable, before disassembling a complicated chandelier for cleaning or repair, to photograph it for record.

35. The Daughters of the American Revolution Museum example is in the Missouri State Room, and the Missouri Historical Society's chandelier is in a permanent display representing the Ladies' cabin of a Mississippi steamboat.

36. Edwin T. Freedley, *Philadelphia and Its Manufactures* (Philadelphia, Ed. Young, [1859]), p. 355.

37. J. B. Chandler, *Description of the Establishment of Cornelius and Baker, Manufacturers of Lamps, Chandeliers and Gas Fixtures, Philadelphia* (Philadelphia, [1860]), p. 20. Was the reference to the Kremlin merely advertising hyperbole or based on fact? It has not been possible to answer that question. The architect Konstantin Andreevich Ton completed the Great Palace in the Moscow Kremlin for Tsar Nicholas I in 1849. Because the monarch esteemed George Washington Whistler (1800-1849), the American engineer who built the Saint Petersburg-Moscow Railway, so much that he honored him with the Order of St. Anne in 1847, it may be possible that he admired American industry enough to order some of his new palace chandeliers from Cornelius and Company. *Dictionary of American Biography,* 1936 ed., s.v. "George Washington Whistler."

38. Starr, Fellows and Company's Illustrated Catalogue, plate 31. The figure numbered 700 has branches identical with those on several chandeliers at Quarters One, Springfield Armory, as seen in 1916 photographs. Craig Littlewood, a craftsman experienced with Cornelius castings, believes the morning-glory branch is a Cornelius design.

39. The Metropolitan Museum of Art, *19th-Century America—Furniture and Other Decorative Arts* (New York: New York Graphic Society, Ltd., 1970), item 113. The chandelier is illustrated in color. Charles Lockwood, *Brick and Brownstone* (New York: McGraw-Hill, 1972), p. 204. Illustrates similar glass elements on a chandelier fitted for eight gas candles.

40. A. D. Jones, *The Illustrated American Biography; Containing Correct Portraits and Brief Notices of the Principal Actors in American History,* vol. 2 (New York: J. Milton Emerson and Company, 1854), pp. 406, 413.

41. *The American Gas-Light Journal* (June 15, 1863), pp. 370-373. List of 433 gas companies in the United States with the dates of their charters. See Appendix.

42. Archer and Warner, *A Familiar Treatise on Candles, Lamps and Gas Lights; with Incidental Matters, Prepared for the Use of their Customers by Archer & Warner, Manufacturers of Gas Fixtures, Chandeliers, Lamps, Girandoles, &c.* (Philadelphia, [1850]), *passim.* Freedley, *Philadelphia and Its Manufactures,* p. 438.

43. Viggo Rambusch, "Technical article: lighting for restorations," *Interiors* (April 1975), p. 172.

44. Ibid.

45. Craig Littlewood, who has restored many gas fixtures and lamps, has observed this to be the case.

46. The Metropolitan Museum of Art, *19th-Century America—Furniture,* items 131, 134, and 135. The cover shows the bronzes in color with other details.

47. Southwestern Bell Telephone Company, *A St. Louis Heritage: Six Historic Homes,* (St. Louis: 1967), pp. 26-27. The Campbell House parlor is illustrated in color. Richard Hubbard Howland, "Tuscan Transplant," *Arts in Virginia* (1968), 9:6.

48. *Art-Journal Illustrated Catalogue* (1851), p. 212.

49. Benjamin Silliman, Jr., and Charles R. Goodrich, eds., *The World of Science, Art, and Industry Illustrated from Examples in the New York Exhibition, 1853-1854* (New York: G. P. Putnam and Company, 1854), p. 157.

50. *Ibid.*, p. 158.

51. Greeley, *Great Industries*, p. 315. Pyne Press, *Lamps*, p. 17.

52. *McElroy's Philadelphia Directory*, 1853.

53. *Boyd's Philadelphia City Business Directory*, 1861.

54. Freedley, *Philadelphia and Its Manufactures*, p. 352, plate opp. p. 35.

55. Federal Writers' Project, *Washington City and Capital* (Washington, D. C.: Government Printing Office, 1937), p. 655.

56. *Frank Leslie's Illustrated Newspaper* (1863), pp. 42-43.

57. Cornelius and Baker, *Description*, p. 24; Freedley, *Philadelphia and Its Manufactures*, p. 356.

58. Nicholas B. Wainwright, *Philadelphia in the Romantic Age of Lithography* (Philadelphia: The Historical Society of Pennsylvania, 1958), pp. 110-112.

59. Jones, *Illustrated American Biography* 2(1854):401.

60. This print has been frequently illustrated: e.g., E. Douglas Branch, *The Sentimental Years 1836-1860* (New York: D. Appleton-Century Company, 1934), opp. p. 252; Davidson, *Life in America* 2:40; Kouwenhoven, *Columbia Historical Portrait of New York*, p. 225; and Harold L. Peterson, *Americans at Home from the Colonists to the Late Victorians* (New York: Charles Scribner's Sons, 1971), plate 197.

61. Kouwenhoven, *Columbia Historical Portrait*, p. 282.

62. Wainwright, *Philadelphia in the Romantic Age*, pp. 87, 153. This ornate store, designed by John Fraser, had a two-story showroom and measured 55 feet wide by 175 feet deep.

63. Jones, *Illustrated American Biography* 2(1854):81, 121, 125.

64. Reproduced in Kouwenhoven, *Columbia Historical Portrait*, p. 273.

65. This watercolor, now in the Bertram K. and Nina Fletcher Little Collection, has been reproduced in Peterson, *Americans at Home*, plate 91.

66. The Metropolitan Museum of Art, *19th-Century America—Furniture*, item 112. A color illustration showing this lamp in a period room installation appeared in *Antiques* (September 1970), p. 405.

67. *Doggett's New York City Business Directory; Wilson's Business Directory of New York City; The New York City Mercantile Register* (1848-1849). The *Mercantile Register* carried an advertisement stating that Johnson's Gas Fittings and General Brass Works made gas pipes and fittings, plain and fancy brass tubing, pendants, brackets, chandeliers, brass bedsteads, gas burners, and cocks. Among the well-known firms listed by Wilson during the 1850s were: 1) Archer, Warner and Company, 376 Broadway; 2) John Cox and Company, 349 Broadway "Importers and dealers in French, English and American gas fixtures;" 3) H. Dardonville, 445 Broadway "Importer of French gas fixtures;" 4) (1854) Mitchell, Bailey and Company; 5) Ringuet Leprince, Marcotte and Company; 6) (1855) Tiffany and Company, 550 Broadway; 7) (1858) Mitchell, Vance and Company; 8) Warner, Peck and Company; and, of course, 9) (1859) Starr, Fellows and Fellows, Hoffman and Company. Wilson's *Business Directory* for 1860 listed among the 24 gas fixture makers the following: 1) Archer, Pancoast and Company, wholesale manufacturers, 9 Mercer Street; 2) Fellows, Hoffman and Company; 3) E. V. Haughwout and Company, 488 Broadway; 4) Isaac P. Frink, 104 Worth Street; 5) Mitchell, Vance and Company, 620 Broadway and 339 W. 24th Street; 6) Tiffany and Company; and 7) Warner, Peck and Company. Note that several firms listed themselves as importers of French or English fixtures. E. V. Haughwout's billhead (reproduced in Kouwenhoven, *Columbia Historical Portrait*, p. 245)

characterized the firm as "wholesale and retail dealers in Cornelius & Baker's chandeliers & gas fixtures," among other items. The fine cast-iron store erected by Haughwout in 1857 at Broadway and Broome Streets still stands. Tiffany dealt in more than jewelry; Ringuet Leprince, Marcotte and Company was a leading decorating and furniture making firm of New York.

68. A circular letter dated March 2, 1857, pasted to the endpaper of the Starr, Fellows catalogue says,

> The co-partnership heretofore existing under the name and style of Starr, Fellows & Co. expired by limitation on the first of February, 1857. Wm. H. Starr, so long and so favorably known as the head of the house having sold his interest therein to the remaining partner and they having associated with themselves Mr. J. A. G. Comstock, the business will be continued as heretofore, at the same locality, under the name and style of Fellows, Hoffman & Co. . . .

In 1847, William H. Starr, lamps, was listed at 67 Beekman Street, and in 1854 Starr, Fellows and Company was listed at the same address with a notation that the address after August 1 would be 74 Beekman Street. This was the locale of Starr, Fellows, and from 1857 through 1870, Fellows, Hoffman and Company. (*Doggett's New York City Directory; Wilson's Business Directory of New York City*). The co-partnership was first composed of William H. Starr, Charles H. Fellows, Charles O. Hoffman, James G. Dolbeare, and George Nichols. Upon the withdrawal of Starr in 1857, the partners were Fellows, Hoffman, Jeremiah A. G. Comstock, Dolbeare, and Nichols. In 1876 the firm was listed at 631-633 Broadway and advertised as "manufacturers of gas fixtures and importers of French bronzes, crystal gas fixtures, French clocks, statuettes, &c." By 1881, they were listed as C. H. Fellows, Hoffman and Company, gas fixtures, at 206 Canal Street, evidently a descent from the eminence of Broadway, and no later listings appeared. (*Wilson's New York City Co-Partnership Directory; Trow's New York City Directory*).

69. The catalogue now in the library of Old Sturbridge Village, Sturbridge, Massachusetts, has about 50 lithographed plates, half of them illustrating gas fixtures. The others illustrate oil lamps. The gilt-embossed cloth cover of the quarto volume shows an oil-lamp chandelier. A few fragments of what appears to have been an Archer and Warner catalogue of the early 1850s have been found attached to correspondence in the National Archives (R.G. 121) relating to the furnishing of the U. S. Custom House in Wheeling, West Virginia.

70. The word "slides" refers to cork-sealed, sleeved stem pipes that could be extended to allow the adjustment of the pendant or chandelier to any height desired.

71. Illustrated in *Antiques* (July 1970), p. 96.

72. Because two other Whittemore chandeliers had Starr, Fellows branches, it is possible, unless the supplier dealt with two or more manufacturers, that the attribution should be to the New York firm rather than to Cornelius and Baker.

73. Illustrated in William Seale, *The Tasteful Interlude: American Interiors through the Camera's Eye, 1860-1917* (New York: Praeger Publishers, 1975), p. 137, plate 112.

74. Illustrated in *Antiques* (May 1976), p. 1034. The globes belong properly to lamps, not to a gaselier.

75. *Op. cit.*, Silliman and Goodrich, p. 158.

76. *McElroy's Philadelphia Directory*, 1841-1872; *O'Brien's Philadelphia Wholesale Business Directory*, 1843-1857; *Boyd's Philadelphia City Business Directory*, 1859-1860; *Cohen's Philadelphia City Directory*, 1860. An attestation dated August 25, 1859, giving the precise dates of the Archer, Warner, and Miskey partnership, is on file in the Office of the Architect of the Capitol, Washington, D. C.

77. *Wilson's Business Directory of New York City*, 1853-1860; *Wilson's Co-Partnership Directory*, 1857-1861; *Trow's New York City Directory*, 1861-1902.

78. The Metropolitan Museum of Art, *19th-Century America—Furniture*, item 103; Denys Peter Myers, *Maine Catalogue—The Historic Architecture of Maine* (Augusta: The Maine

State Museum, 1974), pp. 121-122. If the J. J. Brown House was equipped with gas when built in 1845, the attribution should be to Archer solely, as Warner did not become a partner until 1848. It should be noted that the Portland Gas-Light Company was not chartered until 1849, but it may well have begun operations, like several other gas companies, before receiving its charter.

79. Freedley, *Philadelphia and Its Manufactures*, p. 438.

80. *Ibid.*, p. 439.

81. *Ibid.* The railing is illustrated in Myrtle Cheney Murdock, *Constantino Brumidi—Michelangelo of the United States Capitol* (Washington: Monumental Press, Inc., 1950), p. 9.

82. Thomas Webster and Frances Byerley Parkes, *An Encyclopaedia of Domestic Economy* (New York: Harper and Brothers, 1849), p. 202, para. 764. The first edition appeared in London in 1844, the year of Goodyear's patent (see footnote 19).

83. It has not been considered necessary to discuss governor burners in detail as their use in America was so limited. William Sugg, the English manufacturer of gas equipment, developed a number of such devices, the best of which was probably one introduced around 1880 that used a steatite float. As early as 1867, Julius Bronner of Frankfurt-am-Main produced a governor burner with a steatite plug, and one Giroud developed a patent governor called a "Rheometer" in 1871 that was improved by one Peebles in 1875. Chandler, *Outline*, pp. 91-95.

84. The Oertel painting is illustrated in Peterson, *Americans at Home*, plate 126. Plate 167 in Peterson reproduces half of a D. R. Holmes stereograph taken around 1875 of Senator Charles Sumner's study. A gaslamp attached by a hose to a chandelier may be clearly seen.

85. An interior photograph of the Leland Stanford House of 1869-1871 in Sacramento, California, shows a six-branched chandelier with an additional central burner that could be lowered by what appears to be Monson's patent device. *See* Seale, *Tasteful Interlude*, p. 47. ill. 21.

86. G. and D. Cook and Company's *Illustrated Catalogue of Carriages and Special Business Advertiser* (New Haven: Baker and Godwin, 1860; reprinted ed., New York: Dover Publications, Inc., 1970), p. 138. Dover Publications reprinted the Cook catalogue in 1970 as part of their Dover Pictorial Archive Series. Because the patent was dated ca. 1860, and because the Civil War brought construction by the Treasury Department to a halt until well after Bowman was transferred to other duties, it is uncertain whether many Monson patent equipped fixtures were actually used in Federal buildings.

87. William Paul Gerhard, *The American Practice of Gas Piping and Gas Lighting in Buildings* (New York: McGraw Publishing Company, 1908), p. 146.

88. Winslow Ames, "The Vermont Statehouse and Its Furniture," in *Antiques* (August 1965), pp. 200-204; Cornelius and Baker, *Description*, p. 24.

89. Ames, "Vermont Statehouse."

90. Cornelius and Baker, *Description*, pp. 23-24.

91. Freedley, *Philadelphia and Its Manufactures*, p. 355.

92. Mario E. Campioli, "Building the Capitol," in Charles E. Peterson, ed., *Building Early America—Contributions Toward the History of a Great Industry* (Radnor, Pennsylvania: Chilton Book Company, 1976), p. 227.

93. Freedley, *Philadelphia and Its Manufactures*, p. 356.

94. Data sheet dated June 14, 1968, accompanying negative #12567 on file in the Office of the Architect of the Capitol.

95. "The California Trail" in Thomas Froncek, "Winterkill, 1846," in *American Heritage* (December 1976), pp. 35-41. Additional data supplied by Donald J. Lehman.

96. Washington, D. C., National Archives, R.G. 121. Records of the Public Buildings Service, Office of the Supervising Architect, "Photographs of Designs, Bureau of Construction, Treasury Department."

97. The Hawthorne painting is reproduced in color in *American Heritage* (February 1971), pp. 30-31.

98. Peterson, *Americans at Home*, plate 200. This plate reproduces the Hawthorne in black and white. Peterson mistakes the striped flooring not uncommon in the 1860s, for floor cloth and incorrectly surmises that the beer pumps dispensed "coffee or other non-alcoholic beverages."

99. The small hook shown on plate 60 above the burner was provided for the purpose of hanging a smoke bell.

100. The Stanton Hall bronze fixtures are alleged to have been imported from France. However, no documentary evidence has been produced to support the allegation, except that the term "French bronze" has long been associated with the fixtures. Against the contention that they were imported may be cited the fact that "French bronze" was a common trade term used to designate a particular finish. Furthermore, no French fixtures are known that resemble the Natchez examples in style, whereas many examples of the Philadelphia school resemble them closely. All the allegorical subjects on them that can be identified refer to American history. It may be observed that several marked fixtures by Cornelius and Baker are extant in Natchez.

101. *The Art-Journal; The Illustrated Catalogue of the Universal Exhibition Published with The Art Journal* (London: Virtue and Company, [1867]), p. 139.

102. Treasury Department data were supplied by Donald J. Lehman.

103. Unfortunately, despite their promise, the editors of *The Art Journal* did not mention the "interesting process" elsewhere in the *Illustrated Catalogue*.

104. The publication in 1868 in London of Eastlake's *Hints on Household Taste* spread his ideas like wildfire. Even before the first American edition of his influential work appeared in 1872, arbiters of taste in this country were pressing for the adoption of his rather ascetic aesthetic principles.

105. Jay E. Cantor, "A Monument of Trade — A. T. Stewart and the Rise of the Millionaire's Mansion in New York," *Winterthur Portfolio* 10 (Charlottesville, Virginia: University Press of Virginia, 1975), pp. 165-197. Cantor's article discusses both the aesthetic and the sociological significance of the house. The photographs of the interiors shown here on plates 66 and 67 are reproduced from *Artistic Houses* (New York: D. Appleton and Company, 1883-1884), vol. 1, part 1. The difference in design concept between the Stewart drawing room fixtures of 1869 and the Morse-Libby House music room chandelier of 1863 (plate 62) is obvious when the two plates are compared. Clearly, a new trend was setting in. Elm Park, the Lockwood-Matthews Mansion of 1868 in Norwalk, Connecticut, probably the most lavish American country house of its time, originally had gas fixtures similar to the Stewart chandeliers, but they were lighter in design. *See* Seale, *Tasteful Interlude*, pp. 43-45.

106. For examples of the type, see plates 73, 83, 84, 85, 89.

107. Cantor, "Monument of Trade," p. 187, fig. 22.

108. Dean Hale, "Diary of an Industry." See also note 18.

109. *Cook and Company's Illustrated Catalogue of Carriages*, p. 106. The St. Nicholas Hotel advertisement quotes the *New York Pathfinder* at length and illustrates the hotel's main dining room lighted by three large chandeliers and numerous brackets.

110. Advertising "Letter" dated January 19, 1856, reproduced in Leslie Dorsey Janice Devine, *Fare Thee Well* (New York: Crown Publishing Inc., 1964), pp. 64-65. Possibly the source of the St. Nicholas Hotel lighting system was the Ornamental Iron Works of Philip Tabb

at 522 Broadway "opposite St. Nicholas Hotel." Tabb advertised "Portable Gas Works made to order. Contracts taken for building Gas and Water Works. . ." in 1860.

111. Roy T. Bramson, *Highlights in the History of American Mass Production* (Detroit: Bramson Publishing Company, 1945), p. 52.

112. Jones, *Illustrated American Biography*, 2(1854):393. After the development of oil drilling (for petroleum) at Titusville, Pennsylvania, in 1859 by Edwin L. Drake (1819-1880) who appears to have had no connection with O. P. Drake, gasoline replaced benzine as the agent for vaporization in gas machines.

113. *American Gas-Light Journal* (December 1, 1859), p. 103.

114. [Springfield Gas Machine Company], *Circular of the Springfield Gas Machine Co. of Springfield, Mass. Manufacturers of Portable Gas Machines and Contractors for the Erection of Gas Works, Suitable for the Lighting of Mills, Factories, Machine Shops, Hotels, Public Halls, Churches, Blocks of Stores, Private Dwellings, or Any Class of Buildings Beyond the Reach of Coal Gas Mains. Also, Manufacturers of Carbureting Apparatus for the Purpose of Enriching Coal Gas. Also, Dealers in Gasoline for Gas Machines* (Springfield: [Samuel Bowles and Company, 1867]). Testimonials in this eight-page pamphlet are dated 1866 and refer to the satisfactory use of "vapor of Naphtha" (gasoline) in the previous year. *Poor Richard's Gas Catechism for the People* (Springfield, 1870), p. 15 refers to gasoline as the substance from which gas "is made" in portable gas machines. That pamphlet was probably issued by the Springfield Gas Machine Company. The status of the Springfield Gas Machine is indicated by the fact that a leading technical publication of the period illustrated and described that particular device for the major part of its article titled "Gas, Illuminating, Machines for Producing." *Appleton's Cyclopaedia of Applied Mechanics: A Dictionary of Mechanical Engineering and the Mechanical Arts* (New York: D. Appleton and Company, 1880) 1:935-938.

115. The dotted lines running between the house cellar and the buried generator at the left are labeled "Gas Pipe" and "Air Pipe."

116. The Guy painting was reproduced in color in *American Heritage* (April 1966), pp. 8-9. The Johnson painting was reproduced in color in Metropolitan Museum of Art, *19th-Century America—Paintings and Sculpture* (New York: New York Graphic Society Ltd., 1970), item 144, and in *American Heritage* (October 1966), p. 53. A lithograph published in 1865 by *Leslie's Chimney Corner* shows the east room of the White House gaslighted during President Lincoln's Inauguration Day reception. At the other end of the social spectrum, a painting entitled "An Evening at the Ark" done by Julius Gollmann in 1859, now in the Western Reserve Historical Society in Cleveland, shows a cheaply furnished room papered in arsenical green wallpaper during a meeting attended by a group of poorly dressed men. The stark interior is lighted by the single fishtail burner of a plain pendant.

117. The description referred to the steamers "Bristol" and "Providence," built in 1867. Old Colony Railroad, *The Popular Resorts of Massachusetts and Newport, R. I.* (Boston: Old Colony Railroad, 1878).

118. Gerhard, *American Practice of Gas Piping*, p. 146.

119. The Metropolitan Museum of Art, *19th-Century America—Furniture*, item 161.

120. Chandeliers fitted with gas candles are among the types *not* included in the Mount Washington Glass Works display. The firm made shades as well as fixtures and evidently wished to encourage their use. Their advertisements show that they were particularly proud of their painted shades. During the 1880s Louis XVI Revival crystal chandeliers fitted with gas candles were used in some fine mansions. An excellent matching set composed of a chandelier and brackets in that style was in the drawing room of the Oliver Ames, Jr., House (1882) at 355 Commonwealth Avenue in Boston and is illustrated in Seale, *Tasteful Interlude*, pp. 74-75. "[Mount Washington Glass Works] is the only factory in the country where crystal chandeliers are made complete." New Bedford Board of Trade, *History of New Bedford* (New Bedford: Board of Trade, 1889).

121. The Metropolitan Museum of Art, *19th-Century America—Furniture*, items 176, 177. The chandelier is illustrated in color in *Antiques* (September 1970), p. 409.

122. The 1875 firm also included: Samuel Vance, Vice-President; Edgar M. Smith, Secretary-Treasurer; and Edward A. Mitchell, and Dennis C. Wilcox, Trustees with these officers. *Wilson's New York City Co-Partnership Directory*, *Wilson's Business Directory of New York City*, and *Trow's New York City Directory*, *passim*. Mitchell, Vance and Company, *Centennial Catalogue of Chandeliers, Gas Fixtures, Bronze Ornaments, Clocks, Etc.* (New York: 1876), *passim*.

123. *Trow's New York City Directory*, 1881.

124. Mitchell, Vance and Company, *Centennial Catalogue*, unpaginated.

125. Other commercial structures included the Equitable Life Assurance Company building in Boston and the Lord and Taylor Store in New York. For the contemporary importance and historic significance of the Tribune and Western Union Telegraph Buildings, see Winston Weisman, "New York and the Problem of the First Skyscraper" in *Journal of the Society of Architectural Historians* 12 (March 1953): 13-21. As Richardson specified Mitchell, Vance and Company fixtures for his Brattle Square Church, it is quite possible that the great corona of 1877 that once gave unity and scale to the crossing of his Trinity Church, Boston was by the same firm. The fixture is illustrated in Van Rensselaer, Mariana Griswold, *Henry Hobson Richardson and His Works* (1888; reprint ed., New York: Dover Publications, Inc., 1969), opp. p. 61. The removal of that fixture during "improvements" made in the 1930s was misguided. Fortunately, Mitchell and Vance's somewhat similar corona in Sanders Theatre at Harvard's Memorial Hall in Cambridge still exists.

126. As described further in the *Art Journal*, August, 1875: "The chandelier is massive in appearance, but graceful withal, and is finished in . . . verd-antique, and relieved at prominent points by judicious gilding . . . [it] is one of the most elaborate designs of the kind ever executed in this country. The drawings were made by Mr. Charles C. Perring, Chief designer for the company."
Several of the fixtures illustrated in the Mitchell, Vance and Company *Centennial Catalogue* are also illustrated in more readily available sources, e.g., *A Facsimile of Frank Leslie's Illustrated Historical Register of the Centennial Exposition 1876* (New York: Paddington Press, Ltd., 1974), p. 306 (accompanying text on p. 296) and *Asher and Adams' Pictorial Album of American Industry 1876* (1876; reprint ed., New York: Rutledge Books, 1976), p. 133. Unfortunately, nothing has been discovered concerning Charles C. Perring. He does not appear in any of the standard published reference sources.

127. See plates 15, 27, 53, and 64.

128. *Gopsill's Philadelphia City Directory*, 1887, p. 375.

129. *McElroy's Philadelphia Directory*, 1841-1872.

130. Cornelius and Company issued a 13 page catalogue in 1877 with the title *Examples of Gas Fixtures and other Metal work for Ecclesiastical and Domestic Use. Designed after the Manner of Medieval Art Works by J. M. Beesley.*

131. *Illustrated London News*, May 21, 1853.

132. For the medieval prototype, see Kenneth John Conant, *Carolingian and Romanesque Architecture 800 to 1200* (Baltimore: Penguin Books, 1959), p. 16, fig. 3 (A).

133. *Asher and Adams' Pictorial Album*, p. 49. The article describes the processes carried on at Archer and Pancoast's manufactory and also notes that:
we import bronzes—real and imitation—but only to a very limited extent, as compared with the quantity manufactured at home. Since 1860, the bronze manufacture—the most important feature of which is the production of gas fixtures, has greatly increased in importance. The development of the zinc mines of Lehigh Valley, Pa., and the late discoveries of spelter, as zinc is called in the trade, in New Jersey, Illinois and Missouri, have made American manufacturers altogether independent of foreign mines, and they now turn out goods with which foreign manufacturers cannot successfully compete.

134. Wilson H. Faude, "Associated Artists and the American Renaissance in the Decorative Arts," *Winterthur Portfolio* 10 (Charlottesville, Virginia: University Press of Virginia, 1975), p. 123, fig. 23; Wilson H. Faude, "Mark Twain's House in Hartford, Connecticut," *Antiques* (October 1974), p. 636, fig. 3.

135. *The Art Journal* 2(1876):56.

136. Possibly J. F. Travis was the firm's principal designer, comparable to Charles C. Perring at Mitchell, Vance and Company. Unfortunately, no information about Travis has come to light; he is not listed in any standard reference work.

137. For earlier uses of gas candles, see plates 15 and 71, and note 21 of this report. The Morse-Libby House of 1863 in Portland, Maine, has dining room brackets with gas candles, although the chandelier has shaded burners. The James M. Beebe House ca. 1865 in Boston had a dining room chandelier with a center oil lamp and real candles, but gas candles in the wall brackets. Seale, *Tasteful Interlude*, p. 68, fig. 42. The dining room of the George Finch House in St. Paul had a chandelier dating from around 1880 to 1885 in the Anglo-Japanese taste that was fitted with gas candles. Seale, *Tasteful Interlude*, pp. 94-95, fig. 69.

138. As early as 1844 Petit's Shawl Store in Boston was reported to have single-paned plate glass windows, each containing 48 square feet; and A. T. Stewart's store in New York had French plate glass windows measuring 7 feet wide by 11 feet 2 inches high, or 77 square feet. In 1853, Taylor's Saloon in New York had windows of plate glass 7 feet wide by 16½ feet high. The fact that these were reported in the newspapers indicates that they were exceptional. It was not until 1853 or 1854 that an attempt was made to make plate glass in America. Kenneth M. Wilson, "Window Glass in America," in Charles E. Peterson, ed., *Building Early America—Contributions Toward the History of a Great Industry* (Radnor, Pennsylvania: Chilton Book Company, 1976), pp. 161-164.

139. Isaac P. Frink, *Frink's Patent Reflectors* [New York: 1883]. Patents were issued to Frink on April 10, April 17, and June 12, 1860; December 24, 1861; June 8 and July 7, 1869; February 8, 1870 (patents #3826 and #3827); April 9, 1872; April 17, 1874; May 20, 1879; January 10, 1882; and April 3, 1883.

140. The fixture is illustrated in color in *Antiques* (October 1974), p. 639.

141. Washington, D.C., National Archives, R.G. 42, case 2, folder 9. Richard von Ezdorf was born in the Palazzo Balbi in Venice on the eve of the Revolution of 1848, during which his family suffered much for its loyalty to the Austrian crown. The young aristocrat experienced a particularly bad year in 1866. On July 24 he was wounded at the second Battle of Custozza, and his family sustained grave financial losses after August 23, when Austria signed the Treaty of Prague ceding Venetia to Italy at the end of the Seven Weeks' War. The misfortunes of 1866 may well account for Ezdorf's sailing for New York in 1872 after completing his studies at the Technische Hochschule in Stuttgart, the universities at Innsbruck and Graz, and the Academia del' Arte in Venice. In 1873 he placed his education in architecture, engineering, and the fine arts at the service of the U.S. government. He first appeared on the payroll of the State Department wing of the new State, War, and Navy Building and divided his time between it and other work in the Supervising Architect's office. From 1876 until late in 1886, he was on the War Department rolls while devoting his full time to the State, War, and Navy Building, where the interior ornaments and exterior sculpture were executed from his designs. Thereafter, he was with the Supervising Architect's office again until 1898. He then worked for the Navy Department until 1920, when he retired after more than 47 years in Federal service. He died in 1926. Richard von Ezdorf was an exceptionally gifted designer for his time and a delineator of truly superior talent. Data on von Ezdorf are based on Donald M. Lehman, *Executive Office Building*, General Services Administration Historical Study No. 3, (Washington: Government Printing Office, 1970), pp. 46-53, 83.

142. Lehman, *Executive Office Building*, p. 49.

143. Washington, D.C., National Archives, R.G. 42.

144. Washington, D.C., National Archives, R.G. 77, Records of the Office of the Chief of Engineers, Fortifications File, 1776-1920, drawer 156, box no. 9.

145. By 1886 small-necked shades were already called "old-fashioned globes." C. J. Russell Humphreys, *Gas as a Source of Light, Heat and Power* (New York: A. M. Callender and Company, 1886), pp. 11-12, figs. 11-12.

146. *McElroy's Philadelphia Directory* and *Gopsill's Philadelphia City Directory*.

147. Interior photographs showing the Horticultural Hall chandeliers are reproduced in Robert C. Post, ed., *1876: A Centennial Exhibition* (Washington: The National Museum of History and Technology, Smithsonian Institution, 1976), pp. 66-67 and 72. *Facsimile of Frank Leslie's*, p. 82 has a wood engraving showing the fixtures. A trade card of Thackera, Buck and Company shows Horticultural Hall and says, "Gas fixtures in this building manufactured by Thackera, Buck and Co." Bella C. Landauer Collection, New York Historical Society, vol. 8A.

148. The Metropolitan Museum of Art, *19th-Century America—Furniture*, item 226.

149. Sir William Robert Graves had demonstrated an incandescent electric light as early as 1840, and Sir Joseph Wilson Swan developed a carbon filament electric light bulb in 1860.

150. Chandler, *Outline of History of Lighting*, p. 194.

151. *Financial News*, March 21, 1887. Quoted in Chandler, *Outline of History of Lighting*, pp. 193-194.

152. *Machinery Market*, April 1, 1887. Quoted in Chandler, *Outline of History of Lighting*, p. 194.

153. The advertisement is reproduced in Chandler on p. 200. The statistics quoted are on p. 201. The chapters entitled "Evolution of the Incandescence System of Gas Lighting" and "Development of Incandescence Gas Burners" in Chandler, pp. 179-220, are singularly detailed and excellently illustrated.

154. The photograph of the Scott Studio is reproduced in David A. Hanks, "Isaac E. Scott, Craftsman and Designer," in *Antiques* (June 1974), p. 1312. Clay Lancaster, *New York Interiors at the Turn of the Century* (New York: Dover Publications, 1976), plate 108.

155. The passage continues: "Quite recently, such lamps have been used in connection with the compressed-gas system for the illumination of railroad cars, both here and abroad, and the mantles seem to be but little affected by the vibration and jarring of the cars." Gerhard, *American Practice of Gas Piping*, p. 124.

156. Chandler, *Outline of History of Lighting*, p. 229.

157. Portions of the catalogue are reproduced in Larry Freeman, *New Light on Old Lamps* (Watkins Glen, New York: Century House, 1968) pp. 164-174. Prices included artificial candles and bobeches, but glassware, i.e., shades, was extra. The art glass domes and "seed bead" fringe of two fixtures called "Colonial" were included in the prices. The lamps, or "gas portables," could be had with Argand or Welsbach burners at extra charge. Freeman's *New Light* also reproduces 100 turn-of-the-century shades of different fancy patterns on pages 176-179. The Phoenix Glass Company made 40 of the shades. This firm should not be confused with the earlier Phoenix Glass Works that failed on May 1, 1870.

158. See plate 110, below for a lighter of this type.

159. "Gas-Lighting by Electricity" in *The American Gas-Light Journal* (October 1, 1859).

160. Mario E. Campioli, "Building the Capitol," in Charles E. Peterson, ed., *Building Early America*, p. 227, fig. 12.28.

161. "Gas-Lighting by Electricity" in *American Gas-Light Journal* (October 1, 1859). *Journal of the Franklin Institute*, 3rd ser. (1860) 39:283, 385-389.

162. *Journal of the Franklin Institute*, 3rd ser. (1861) 12:349.

163. Lois B. McCauley, *Maryland Historical Prints 1752 to 1889, A Selection from the Robert G. Merrick Collection Maryland Historical Society and other Maryland Collections* (Baltimore: Maryland Historical Society, 1975), p. 112.

164. The title page of *Morris, Tasker and Co.'s Illustrated Catalogue*, reproduced in Diana S. Waite, *Architectural Elements: The Technological Revolution* (New York: Bonanza Books, n.d.), shows the counterweights and pulleys of a gasholder more clearly than the Hoen lithograph does. Robert M. Vogel, ed., *A Report of the Mohawk-Hudson Area Survey* (Washington: Smithsonian Institution Press, 1973), p. 46, reproduces a Historic American Engineering Record drawing of the Troy Gas Light Company Gasholder House, a domed brick structure built in 1873.

165. Park Benjamin, ed., *Appleton's Cyclopaedia of Applied Mechanics: A Dictionary of Mechanical Engineering and the Mechanical Arts* (New York: D. Appleton and Company, 1880), pp. 900-946. Appleton's very detailed technical article on gas and its manufacture notes that purification with lime was then (1880) almost entirely abandoned in favor of "washers" or "scrubbers."

166. On page 131 of a paper entitled "One of the First Meter Makers in the United States," written for the American Gas Association and published in a now unidentified source, H. C. Slaney, the author, says, "It was in Philadelphia that gas was first produced and exhibited by Michael Ambrosie & Co. at their amphitheatre on Arch Street, between 8th and 9th Sts., in the year 1796." Malcolm Watkins mentions a demonstration of gaslight by one Mr. Henfry in 1799 or 1800 in Baltimore.... (C. Malcolm Watkins, "Artificial Lighting in America: 1830-1860," in *Annual Report of the Board of Regents of the Smithsonian Institution.... 1951* (Washington: Government Printing Office, 1952), p. 393). David Melville of Newport, Rhode Island, first used gas for domestic lighting (in his own house) in 1806 (Watkins) or 1812 (Dean Hale, "Diary of an Industry"). Melville patented his gas machine either in 1810 (Loris S. Russell, *A Heritage of Light: Lamps and Lighting in the Early Canadian Home* (Toronto: University of Toronto Press, 1968), p. 290) or in March 1813 (Hale). In 1813 Melville installed gaslights in a Watertown, Massachusetts, cotton mill and in a mill near Providence, Rhode Island (Hale). On April 23, 1816, Rembrandt Peale demonstrated gaslighting at the Peale Museum in Baltimore, and on June 17, 1816, an ordinance was passed authorizing the Baltimore Gas-Light Company to lay pipes. The first public lamplighting occurred on February 7, 1817 (Hale). For Samuel Morey's use of water gas in 1817, see Alice Doan Hodgson, "History in Towns—Orford, New Hampshire," *Antiques*, October 1977, p. 712.

167. *American Gas-Light Journal* (January 1, 1862), pp. 198-199.

168. *Ibid.*

169. *American Gas-Light Journal* (June 15, 1863), pp. 370-373.

170. *Ibid.* See appendix for a chronology of early gaslight companies in America.

171. Slaney, "One of the First Meter Makers in America," p. 131.

172. *Journal of the Franklin Institute*, 3rd ser., no. 1 (January 1848), 15:337-339.

173. *The American Gas-Light Journal* (November 1, 1859), p. 92. Information on the U.S. Treasury building supplied by Donald J. Lehman, from the U.S. Treasury Voucher, 1852 Accounts.

174. Donald McDonald, *Meters and Meter Makers: A Paper Prepared for the Fiftieth Anniversary Number of the American Gas Light Journal, July 19, 1911* (Albany, New York: C. F. Williams and Son, 1911), p. 22.

175. *Report of the Commissioner of the Patent Office* 1 (1868).

176. A Philadelphia lamp post with eagle finial by Morris, Tasker and Company is among the illustrations on the cover of their catalogue, the second edition of which was issued in 1860. The cover and eight plates from the catalogue are illustrated in Diana S. Waite, *Architectural Elements: The Technological Revolution* (New York: Bonanza Books, n.d.).

177. Information concerning the reconstruction of the lamp in front of Ford's Theatre was supplied by Henry A. Judd, Chief Historical Architect, National Park Service.

178. A photograph dated ca. 1890 of the Dundas-Lippincott Mansion at the northeast corner of Broad and Walnut Streets in Philadelphia shows one of the new type of gas street lamps on that corner, but a short distance from Broad Street on the south side of Walnut Street, one of the old type of lantern-form street lamps is clearly visible. The view is illustrated in Robert F. Looney, *Old Philadelphia in Early Photographs 1839-1914: 215 Prints from the Collection of the Free Library of Philadelphia* (New York: Dover Publications, Inc., 1976), p. 174, plate 168.

179. Henry-Russell Hitchcock and William Seale, *Temples of Democracy: The State Capitols of the U.S.A.* (New York: Harcourt Brace Jovanovich, 1976), p. 119.

180. Walter H. Kilham, *Boston After Bulfinch, An Account of its Architecture, 1800-1900* (Cambridge: Harvard University Press, 1946), p. 51, plate 17.

181. Allan Nevins, ed., *The Diary of Philip Hone 1828-1851* (New York: Dodd, Mead and Company, 1936), p. 302.

Bibliography

Astonishingly little has been written on the subject of gaslighting during the last 50 or 60 years. Compared with the amount of material published on lamps and candlesticks, the material on gas fixtures has been practically nil. It has, therefore, been necessary to include many rare and out-of-print books in this bibliography. This bibliography has been restricted to those works that proved most useful for research. Other publications of more peripheral interest have been included in the notes.

Accum, Fredrick. A *Practical Treatise on Gas-Light, Exhibiting a Summary Description of the Apparatus and Machinery Best Calculated for Illuminating Streets, Houses, and Manufactories with Carburetted Hydrogen, or Coal-Gas; with Remarks on the Utility, Safety, and General Nature of this new Branch of Civil Economy.* London: R. Ackermann, 1815.
This rare and handsomely illustrated work by Friedrich Christian Accum (1768-1838) is the first major publication on gaslighting in any language.

The American Gas-Light Journal, 1859 —
This precursor of the *American Gas Journal* began publication on July 1, 1859, and is the most comprehensive source for technical reports on developments in its field starting from 1859.

Archer and Warner. *A Familiar Treatise on Candles, Lamps and Gas Lights; with Incidental Matters, Prepared for the Use of Their Customers, by Archer and Warner, Manufacturers of Gas Fixtures, Chandeliers, Lamps, Girandoles, &c.* Philadelphia,[1850].
This promotional pamphlet, not a catalogue, is perhaps the earliest extant publication by an American manufacturer of gas fixtures.

Artistic Houses, 2 vols. New York: D. Appleton and Company, 1883-1884.
This work contains many excellent views of American interiors of the post-Civil War era.

Benjamin, Park, ed. *Appletons' Cyclopaedia of Applied Mechanics: A Dictionary of Mechanical Engineering and Mechanical Arts.* New York: D. Appleton and Company, 1880.
Contains one of the best and most complete articles on gas technology of its period in any American publication.

Chandler, Dean. *Outline of History of Lighting by Gas.* [London, South Metropolitan Gas Company, 1936].
Technical and historical information on the development of gaslighting in Great Britain.

Cornelius and Baker. *Description of the Establishment of Cornelius and Baker, Manufacturers of Lamps, Chandeliers and Gas Fixtures, Philadelphia.* Philadelphia: J.B. Chandler, 1860.
An important historical document; it describes the manufacturing processes of the firm in considerable detail.

Cornelius and Company. *Examples of Gas Fixtures and Other Metal Work for Ecclesiastical and Domestic Use. Designed after the Manner of Mediaeval Art Works by J. M. Beesley.* Philadelphia, [1877].
The only known copy of this work is catalogued in the Library of Congress but is presently missing.

Frank Leslie's Illustrated Newspaper, 1855 —
This major weekly is a prime source for pictures of American interiors, particularly interiors of public buildings, and their lighting fixtures from 1855 to about the end of the gaslighting era.

Freedley, Edwin T. *Philadelphia and Its Manufactures: A Hand-Book Exhibiting the Development, Variety, and Statistics of the Manufacturing Industry of Philadelphia in 1857 Together with Sketches of Remarkable Manufactories; and a List of Articles Now Made in Philadelphia.* Philadelphia: Edward Young, 1859.
Contains detailed descriptions of Cornelius and Baker's manufacturing processes and shorter descriptions of Archer and Warner's operations.

Freeman, Larry. *New Light on Old Lamps.* Watkins Glen, N.Y.: Century House, 1968.
Contains useful illustrations from unidentified 1900 gas fixtures and glass catalogues.

Frink, I [saac]. P. *Frink's Patent Reflectors . . .* [New York, 1883].
This catalogue contains useful information on commercial and industrial (as well as some ecclesiastical) lighting during the 1870s and 1880s.

Gerhard, William Paul. *The American Practice of Gas Piping and Gas Lighting in Buildings.* New York: McGraw Publishing Company, 1908.
This is probably the best technical work on the subject published in America after 1900. It includes the Philadelphia rules for gasfitting, regarded as models for the whole country. Gerhard was a civil engineer and a corresponding member of the American Institute of Architects.

Greeley, Horace, et al. *The Great Industries of the United States: Being an Historical Summary of the Origin, Growth, and Perfection of the Chief Industrial Arts of this Country.* Hartford: J.B. Burr and Hyde, 1873.
Contains an excellent article about Cornelius and Sons.

Hale, Dean. "Diary of an Industry," *The American Gas Journal,* 1970. Unpaginated reprint.
A useful chronological series of brief historical notes.

Harper's Weekly, 1857-ca. 1912
The comment following *Frank Leslie's Illustrated Newspaper* applies equally to *Harper's Weekly.*

Humphreys, C. J. Russell. *Gas as a Source of Light, Heat and Power.* New York: A.M. Callender and Co , 1886.
An attractively designed and informative publication of the 1880s written for the layman. A clear, concise source of information.

Journal of the Franklin Institute of the State of Pennsylvania, for the Promotion of the Mechanic Arts. Devoted to Mechanical and Physical Science, Civil Engineering, the Arts and Manufactures, and the Record of Patent Inventions. Philadelphia: Franklin Institute, 1826.
The last words of the lengthy title, "Record of Patent Inventions," are the most significant in this context. The Franklin Institute *Journal* is, for the years before 1858, as important a source of technical information as is the *American Gas-Light Journal* for the years after 1859.

Kouwenhoven, John A. *The Columbia Historical Portrait of New York: An Essay in Graphic History in Honor of the Tricentennial of New York City and the Bicentennial of Columbia University,* Garden City, N.Y.: Doubleday and Company, Inc., 1953.
This excellent source, containing many interior views showing gas fixtures, has recently been reprinted in paperback.

Lancaster, Clay. *New York Interiors at the Turn of the Century in 131 Photographs by Joseph Byron from the Byron Collection of the Museum of the City of New York.* New York: Dover Publications, Inc., 1976.
At least 15 combination gas and electric fixtures, as well as older gas fixtures and numerous electroliers are shown in these photographs taken between 1893 and 1916. Plates 5 and 41 show lighted gas candles. As in Kouwenhoven, many nondomestic interiors are included in this useful visual source.

Larkin, James. *The Practical Brass and Iron Founder's Guide: A Concise Treatise of the Art of Brass Founding, Moulding, Etc. with Numerous Practical Rules, Tables, and Receipts for Gold, Silver, Tin, and Copper Founding; Plumbers, Bronze and Bell Founders, Jewellers, Etc.* Philadelphia: A. Hart, 1853.
Contains some "receipts" for brass finishes that were used during the 1850s.

McDonald, Donald. *Meters and Meter Makers: A Paper Prepared for the Fiftieth Anniversary Number of The American Gas-Light Journal July 19, 1911.* Albany, N.Y.: C.F. Williams and Son, [1911]
Probably the best historical account of American gasmeters and their manufacturers.

McHenry, John. *The Gas Meter and the Apparatus Used in the Manufacture of Coal Gas, Illustrated and Explained. Also Valuable Information Relative to Gas Fittings, Gas Burners and*

the Mode of Burning Gas, the Mode of Estimating the Illuminating Power of Gas, Naphthalized Gas, Ventilation of Gas Lights, the Illuminating Power of Gases, Comparison of Various Methods of Illuminating with Each Other The Cost of Manufacturing Gas at the Philadelphia and Louisville Gas Works, to Which is Added a Brief Sketch of the Early History and Progress of Gas Lighting in England, and Observations on the Generation of Illuminating Gas, Cincinnati: J. Ernst, 1853.
The title says it all. This dull but useful work is a mine of information on American gas practice at mid-century.

McKenney and Waterbury Company. Gas Catalogue G. Boston, [n.d.].
A typical as well as handsome and illustrative example of the many American catalogues dating from ca. 1900.

Metropolitan Museum of Art. 19th-Century America — Furniture and Decorative Arts: An Exhibition in Celebration of the Hundredth Anniversary of the Metropolitan Museum of Art, April 16 through September 7, 1970. New York: New York Graphic Society, Ltd., 1970.
Contains illustrations of and data on eight gas fixtures of high quality.

Metropolitan Museum of Art. Print Room, Elisha Whittlesey Fund 51.533 [Archer and Pancoast Manufacturing Co.]
These 110 chromolithographed plates excellently illustrate this company's best products during the 1870s.

Mitchell, Vance and Company. Centennial Catalogue of Chandeliers, Gas Fixtures, Bronze Ornaments, Clocks, Etc. New York, 1876.
Although it has comparatively few illustrations (wood engravings, not lithographs), this catalogue contains valuable information about the firm, as well as a significant list of references.

Peckston, Thomas S. A Practical Treatise on Gas-Lighting in Which the Gas Apparatus Generally in Use is Explained and Illustrated by Twenty-two Appropriate Plates. London: Herbert, 1841.
Contains the latest technical information at the time when gaslighting was on the threshold of its first real popularity in America. This English work is, as the title says, "appropriately illustrated."

Peterson, Charles E., ed. Building Early America: Contributions Toward the History of a Great Industry. Radnor, Pennsylvania: Chilton Book Company, 1976. Chapter 11, "Early Nineteenth-Century Lighting" by Loris S. Russell.
Contains a brief but recently written sketch of gaslighting.

Peterson, Harold. Americans at Home From the Colonists to the Late Victorians. New York: Charles Scribner's Sons, 1971.
Contains numerous interior views showing gas fixtures, some of great interest.

Philadelphia Museum of Art. Philadelphia: Three Centuries of American Art, Bicentennial Exhibition April 11-October 10, 1976. Philadelphia: Philadelphia Museum of Art, 1976.
Contains excellent entries on Robert Cornelius, the artist designer John Henry Frederick Sachse, and two Cornelius fixtures.

Poor Richard's Gas Catechism for the People. Springfield, Massachusetts, 1870.
A compendium of useful information for the layman in question and answer form. An entertaining period piece of promotional propaganda.

Pyne Press. Lamps and Other Lighting Devices 1850-1906. Princeton: The Pyne Press. 1972.
This work emphasizes lamps, but it fully reproduces the Cornelius and Sons catalogue of ca. 1876. in the Historical Society of Pennsylvania. A useful excerpt from Archer and Warner's Familiar Treatise is also included.

Romaine, Lawrence B. A Guide to American Trade Catalogs 1744-1900. New York: R.R. Bowker Company, 1960.
Somewhat outdated, this book remains an indispensible source for finding catalogues.

Russell, Loris S. A Heritage of Light, Lamps and Lighting in the Early Canadian Home. Toronto: University of Toronto Press, 1968.

Although its subject is Canadian lighting, many of the examples cited were made in the United States. This is one of the few relatively recent works to mention gaslighting at all.

Seale, William. *The Tasteful Interlude: American Interiors through the Camera's Eye, 1860-1917.* New York: Praeger Publishers, 1975.
As far as views of American domestic interiors are concerned, this book does for the later 19th century what Peterson's book does for the entire century. Together with Kouwenhoven's *Columbia Historical Portrait,* Lancaster's *New York Interiors,* and Peterson's *Americans at home,* Seale's *Tasteful Interlude* is one of the four best collections of interior views showing contemporary installations of gas fixtures.

Starr, Fellows and Company. *Starr, Fellows and Company's Illustrated Catalogue of Lamps, Gas Fixtures, &c.* [New York] 1856.
This major document is the first American gas fixture catalogue so far discovered in complete form.

Wainwright, Nicholas B. *Philadelphia in the Romantic Age of Lithography: An Illustrated History of Early Lithography in Philadelphia with a Descriptive List of Philadelphia Scenes Made by Philadelphia Lithographers Before 1866.* Philadelphia: Historical Society of Pennsylvania, 1958,
Contains some excellent views of shop interiors during the 1850s.

Walter, Thomas U [stick] and Smith, J. Jay. *A Guide to Workers in Metal and Stone: For the Use of Architects and Designers, Black and White Smiths, Brass Founders, Gas Fitters, Iron Masters, Plumbers, Silver and Goldsmiths, Stove and Furnace Manufacturers, Pattern Makers, Marble Masons, Stucco Workers, Carvers and Ornamental Workers in Wood, Potters, Etc. From Original Designs and Selections Made from Every Accessible Source, American and European.* Philadelphia: Carey and Hart, 1846.
Most of the gaslights shown in this work are exterior fixtures of English design, but plate 17 is the earliest illustration of American gas fixtures (by Cornelius) presently known.

Warner, Miskey and Merrill. *Patterns.* Philadelphia: P.S. Duval and Son, [1859].
A set of over 40 extremely handsome colored lithographs of Archer, Warner, Miskey & Co. fixtures, 1857-59.

Washington, D.C., Library of Congress, Lot 2728. Division of Prints and Photographs.
Group of 69 British engravings of gas fixtures dates from 1820-1830 and is probably the oldest collection of such illustrations still extant.

Washington, D.C., National Archives, Record Group 77.
Records of the Office of the Chief of Engineers, Fortifications File, 1776-1920. Thackera Sons and Company.
The best group of illustrations of American gas fixtures ca. 1880 that has so far been discovered.

Watkins, C. Malcolm. "Artificial Lighting in America: 1830-1860." *Annual Report of the Regents of the Smithsonian Institution Showing the Operations, Expenditures, and Condition of the Institution for the Year Ended June 30, 1951.* Washington, U.S. Government Printing Office, 1952, pp. 385-407.
A pioneering article with a brief discussion of gaslight.

Webster, T [homas] and Parkes, Mrs. [Frances Byerley]. *An Encyclopaedia of Domestic Economy: Comprising Subjects Connected with the Interests of Every Individual; Such as the Construction of Domestic Edifices; Furniture; Carriages, and Instruments of Domestic Use. Also, Animal and Vegetable Substances Uses as Food, and the Methods of Preserving and Preparing Them by Cooking; Receipts, Etc. Materials Employed in Dress and the Toilet; Business of the Laundry; Preservation of Health; Domestic Medicines, &c.* New York: Harper and Brothers, 1849.
Thomas Webster's *Encyclopaedia* was first published in London in 1844, and this 1849 American edition was entered according to an Act of Congress in 1845. This work is by no means as superficial as the title page seems to suggest. Book four, titled "Artificial Illumination," is well illustrated with British lighting devices, and chapter five of this book, "Illumination by Means of Gas," (pp. 198-206) is an excellent early article on the subject.

Yorke, Eugene H. *The Essential Facts in Lighting with Kerosene, Electricity, Gas.* [n.p.], 1890.
An entertaining promotional brochure favoring gas written for laymen.

Index

House, n9; Tribune Building, 73, n125; Tucker Manufacturing Company, 65; William Henry Vanderbilt House parlor, 69; Western Union Telegraph Company Building, 73, n125; Windsor Hotel, 73

Nichols, George, manufacturer (partner, Starr, Fellows and Co., and Fellows, Hoffman and Co.), n68

Nipples, 44

Norfolk, Virginia, Moses Myers House; 18

North Attleboro, Massachusetts, Edmund Ira Richards House; 26

Northampton, Massachusetts, "Smith's Female College"; 73

Norwalk, Connecticut, Lockwood-Matthews Mansion (Elm Park); n105

Octagonal lantern; 116, 118

Orford, New Hampshire, Samuel Morey House, site of first water gas installation; n166

Osler, F. and C., English glass manufacturers; 73

Ottawa, Canada, Houses of Parliament; 88

Pancoast, George, manufacturer (partner, Archer, Pancoast and Co.); 45

Paper shades; octagonal, n19; reflector, 39

Paris, France; Opera chandelier, 110; Universal Exhibition of 1867, 49, 65

Parlor fixtures; Int., 18-28, 39, 43, 49, 66, 69-71, 73, 75-77, 84, 106, 107, n21, n105; back parlor fixtures, 9, 69, 82

Patents; Amsterdam for "carbonizing" gas, 68; Archer lard lamp, 22; Archer and Pancoast "Excelsior centre-light attachment," 84; Archer and Warner designs, 21; Cornelius (first American patent for a gas jet), 15; Edwards, gas switch, 101; Frink reflectors, 88, n139; Gardiner lighter, 110; Gardiner switch, 110; Giroud governor, n83; Goodwin gasmeter, 112; Lebon wood gas, Int.; Melville gas machine, 68, n166; Meyers lighter, 110; Milne fishtail burner, 3; Mitchell, Vance "double slide centre light chandelier," 73, 84; Monson's extension fixture, 52, n85, n86; Morey water gas apparatus, n166; Neilson fishtail burner, 3; Potts gasmeter, 112; Weber-Marti gas switch, 101; Welsbach mantle, 99; Wilson lighter, 110

Peale, Charles Willson and Rembrandt, early gaslight demonstration; 16, n166

Peck, James B., manufacturer (partner, Archer, Warner and Co., and Warner, Peck and Co.); 45

Peebles, inventor of improved "Rheometer" governor; n83

Pendants (also sp. pendents); 8, 37, 39, 58, 60, 81, 96, 106, n116; term defined, 1; reading, 41

Perring, Charles C., designer for Mitchell, Vance and Co.; 73, n126, n136

Petroleum gas; 111

Philadelphia Gas Company; 15

Philadelphia, Pennsylvania; Academy of Music, 29, 119; Ambrosie's Amphitheatre, n166; Arch Street Theatre, 114; Continental Hotel, 30; Dundas-Lippincott Mansion, n178; early demonstrations of gaslight in, n166; Horticultural Hall, 93, n147; House on Walnut Street, 112; Independence Hall, 16; Independence Square, 117; lampposts in, n176; Lee and Walker Music Store, 119; L. J. Levy and Co. Store, 30; Lewis House, 14; manufacturing in, 33; Maxwell Mansion (Germantown), 18; Oakford's Model Hat Store, 30; Peale's Museum, 16; Second Bank of the United States, 119; Mrs. A. W. Smith, parlor, 32; Wood and Perrot Foundry, 118

Phoenix Glass Company, manufacturers of shades; n157

Pillars; 6, 12, 16, 30, 43, 50, 73, 81; term defined as standing fixture, 1; term used for stem of fixture, 14, 106, 107, 110; on newel post, 92

"Pillar icicles"; 14, n17

Pilot lights; 54, 110

Pipe; appropriate finish of, 40; fixtures made of, 29; rods, 10; service, int.; use in U.S. Capitol, 54; without canopy, 25

Polychromed fixtures; 73, 80, 81

"Portable gas works," *see also* Gas machines; 68, n18, n110, n114

"Portables," *see* Gaslamps

Portland, Maine; John J. Brown House, 46, n78; Morse-Libby House, 62, n105, n137; Portland Gas-Light Company, n78

Potts, Albert, gasmeter patentee; 112

Prisms, see also "Icicles"; 11, 13, 14, 29, 92; added, 54; notched spear, frontispiece, 70, 97

Providence, Rhode Island, Beneficent Congregational Meeting House ("Round Top Church"), 27; mill near, n166

Public building fixtures; 52-60, 65, 73, 88, 90, 91, 93, 108, 109, 118, n147